Meteorology and Climatology

nova
science publishers
www.novapublishers.com

Meteorology and Climatology

Wind Speed: An Overview
Nicolas Kočí (Editor)
2020. ISBN: 978-1-53618-412-9 (Softcover)
2020. ISBN: 978-1-53618-487-7 (eBook)

More information about this series can be found at
https://novapublishers.com/product-category/series/meteorology-and-climatology/

Saeid Eslamian,
Mir Bintul Huda
and Faezeh Eslamian
Editors

Handbook of Uncertainty in Eurasian Forecasting (HEF)

nova
science publishers
www.novapublishers.com

DOI: https://doi.org/10.52305/IWAW5705.

Additional color graphics may be available in the e-book version of this book.

Library of Congress Cataloging-in-Publication Data

ISBN: 978-1-68507-966-6

Published by Nova Science Publishers, Inc. † New York

Contents

Preface

Forecasting is a technique that uses historical data as inputs to make informed estimates that are predictive in determining the direction of future trends. Eurasia comprises about 36% of the world's total area and about 70% of the world population. Eurasia comprises Asia and Europe, although, geographically, it is a single continent with arbitrary geological borders. Eurasia has been a home to the world's oldest civilizations and plays an important part in the mainstream history of the world. Eurasian countries have many common characteristics and forecasting of this region can prove to be of major help in integrated resources management, leading to sustainable development, optimum decision making of international world organizations and achieving goals of world peace. This book deals with the various aspects of social and environmental importance in this region, especially climate change and hydrological modeling and flood forecasting.

Chapter 1 - The aim of this chapter is to present the eponymous and anonymous Saints of Persia as well as a historic religious analysis of the situation in it, mainly in 4th c. CE, bringing the necessary witnesses as a target. The material of this work is monthly books of the Saints of the Christian Church, calendars, the Holy Bible, ancient Greek literature, other historic and religious writings, encyclopedias, journals and electronic pages of the world wide web. The method is the critical analysis of text. Results are the presentation of the Saints by month, the Christians' persecutions in Persia, the presentation of Zoroastrianism and Zoroaster, the adoration of fire, the position of fire in Judaism and Christianity, the reference on account of Magoi – Magians and their martial against Christianity and the short-lived recognition by Persian empire but in its heretical of Nestorianism form. As a conclusion, refers to the inflicted damage of a country from the persecutions of the State on a healthy religious reform.

Chapter 2 - Mental Health Care (MHC) Act 2017 is published in the Gazette by the Government of India. It is aimed to intensify the human rights of persons with mental illness, and it gives them power regarding their health and treatment. Changes such as advance directives and nominated representatives are new, and review boards and responsibilities of government are clearly emphasized. And the act of attempt to suicide was decriminalized which was a big step as it was mentioned as a separate section in the Indian Penal Code. The objective of this Chapter is to give an overview of the Mental Health Care Act 2017 and the upcoming issues, its implementation and the challenges it must face in the coming future.

Chapter 3 - Natural disasters threaten ecosystems, human life, and economic development greatly. However, there is limited research focusing on natural disasters in arid and semi-arid regions of southern Africa. This chapter explains the causes and effects of natural disasters and the disaster management cycle in general. It then zeros into specific major natural disasters occurring in Zimbabwe. It also highlights the framework for understanding natural disasters. The discussion dovetails on the management of natural disasters in the context of civil protection in the country. The Civil Protection Unit is a national disaster response institution that is actively rolling out education and awareness-raising programs so that vulnerable people are more prepared to face disasters and to react adequately when they occur. The Unit, though under-resourced, is actively designing emergency and evacuation plans such as escape routes to safer places in times of flooding.

Chapter 4 - Several anthropogenic activities and climate changes are affecting watershed characteristics. The increased prolonged period between rainfall events is severely affecting the watershed. Groundwater depletion, non-conjunctive use of surface water and groundwater, and increasing water demand with a rising population are stressing a catchment's water demand and supply. Such a deficit in the water supply stresses plant productivity and drought situations in the watershed. For policymakers to address the drought issue and act on probable mitigation methods, it is vital to detect drought conditions and evaluate their intensity at the early stages. It is very challenging for scientists to predict the drought and its severity with better accuracy. Drought monitoring indices help monitor the drought and measure its severity for a particular location. This chapter discusses in detail various drought indices and their uses. Scientists have developed different indices to measure drought over a specific region accurately. Recently, various studies used unequal weightage methods to develop composite drought indices. This chapter discusses several techniques for giving weightage to drought indices to build composite drought indices; various drought indices are developed because they are suitable for a particular location. The suitability of that drought index over other regions is conditional based on rainfall, groundwater availability, and several other factors. The value of drought index may increase or decrease based on the factors that we choose. Policymakers need to choose a location-specific drought index for effective monitoring of drought.

Chapter 5 - The Danube and its tributaries have been crossing mountains and plains in their almost unchanged riverbeds for thousands of years, disregarding national and administrative boundaries. Nevertheless, even decades ago, several countries made only limited data and information available on the water level and flood protection status of their rivers. In recent years, information has changed hands based on bilateral agreements, but for others, information could only be obtained by browsing the Internet, sometimes in a rather complicated way. EU Strategy for the Danube Region Environmental Risks Priority Area initiated a project that is supporting ICPDR in developing the Danube Hydrological Information System. The main aim of the Danube River Basin Enhanced Flood Forecasting Cooperation (DAREFFORT) project is to give a comprehensive overview of the complex national flood and ice forecasting systems and eliminate the shortcomings of the existing forecasting practices as well as improve the exchange and availability of hydrological and meteorological data between the participating countries with establishment of the Danube Hydrological Information System. The Evaluation report on flood and ice forecasting systems and methodologies in the Danube countries produced in the frame of the DAREFFORT project provides more detailed information on flood information services in DRB. In the frame of the DAREFFORT project, all authorized meteorological and hydrological data of the Danube River are stored in a central database and made available online to all licensed hydrological and flood protection institutions for further processing in virtually real-time. At this moment 12 countries of the Danube have joined forces to work out their proposals that are essential for the future.

Chapter 6 - Climate change is an inevitable phenomenon that has led the earth to evolve from an ice age to the present era. Due to the temperature rise, the rate of Evapotranspiration (ET) is increasing which leads to a higher rate (chance) of extreme events (floods). This raises the need to analyze the watersheds which show considerable vulnerability to climate change. This study analyses the effect of environmental (Climate) change on Hathmati River which rises from Gujarat Malwa Hills having a total catchment of 1574 sq. km and average annual rainfall of 864mm. The coefficient of variation of annual rainfall over the basin is rather high and ranges between 42% to 65%. The analysis is classified into two parts 1) Trend Analysis and 2) Hydrological

Modeling. The trend analysis is performed by using Sen's Slope method where significance values for monthly and seasonal data are determined. From trend analysis, it is observed that there exists a considerable positive trend for temperature in accordance with climate change impact. Sen's significance (Z) value ranges from -0.03 to +1.72. The existing trend for rainfall, evapotranspiration & surface runoff is found to be positive with an average significance value of 2.88, 2.07 & 1.59 respectively. The tool used for hydrological modeling is ArcSWAT (semi-distributed hydrological model) which takes in Geo-Spatial data (and meteorological data) to estimate various parameters such as ET, Runoff, etc. The model run has been carried out for 35 years (1979 to 2014) where model outputs are compared with the observed values of runoff. Sensitivity analysis (Calibration) is performed to calibrate the model through the SUFI-2 (using SWAT CUP) algorithm for the monthly values of runoff giving the average efficiency (NSE) of the model as 0.90. The model is successfully validated for five years giving NSE as 0.89. The Calibrated & validated model shows that the average values of evapo-transpiration & surface runoff in mm against 882mm of rainfall are 303mm & 285mm respectively. The analysis has been performed to evaluate the impact of climate change on Evapotranspiration & Surface Runoff through considerable variations in climatic conditions using IPCC scenarios (CMIP-5 projections) and identifying the vulnerable spots that may require attention because of the climate change. The framework is proposed to assess and mitigate the impact of climate change based on the present study.

Chapter 7 - Groundwater sources are very important, particularly in dry areas, and the sustainability of these resources necessitates serious efforts to protect them from contamination. Digital models have become the mainstay of risk assessment studies and groundwater aquifer protection through intrinsic or specific contamination susceptibility models. Groundwater is more likely to harm the consumer population if exposure occurs since fewer measures are also taken to eradicate the contaminant before it reaches the public. For this reason, the conservation of groundwater is of considerable importance, particularly in arid areas with highly vulnerable karstic and fractured basaltic aquifers. Mapping the vulnerability of groundwater contamination has become a key step in groundwater risk assessment and protecting groundwater quality.

This chapter consists of two parts, the first discusses the components of the groundwater risk assessment and examines the evolution of the groundwater vulnerability modeling. The scope of this review involves the design of standard methods for assessing contamination potential, as well as the development of a system that can be easily displayed on maps. Thus, this review is not intended to evaluate the various kinds of groundwater vulnerability modeling. Rather, it is meant to serve as a general base for the variety and evolution of vulnerability evaluation methodologies. The second part explains the visualization of groundwater vulnerability by a systematic study carried out to establish an effective method for assessing the groundwater risk of renewable aquifers in the arid agricultural area by developing a multipurpose database in the GIS setting and validating the overlaying vulnerability method developed by comparing the findings with the groundwater quality characteristics of the area and with former visualization.

Chapter 8 - The rapid prediction of spatial distribution from the near-field tsunami is very important for tsunami risk characteristics. Despite significant improvements in the past decade, physics-based numerical models are still computational. Uncertainty is an important challenge in the risk analysis of tsunamis. Tsunami elevations are affected by complex factors and are continuously changing. Tsunami height has random features. This study suggests that water depth (related to sea topography) is the most important factor affecting the height of the tsunami. But around the world, a significant area of marine topography has not been measured. Therefore, you need to use uncertainty to consider the depth of water. Asia and the Middle East have both volcanic tsunamis and has some popular, most economical, and rapidly developing beaches in the world. This helps examine the risk of tsunamis in Asia and the Middle East.

Chapter 9 - Water is an essential requirement on this earth. But, with urbanization and the population explosion, this resource is under continuous stress. It is evident from the depleting water reserves and the increasing demands of water for various purposes. The changing climate has further put stress on the management practices of dealing with this natural resource that sets apart the earth from all other planets. The chapter focuses on smart water management techniques to reduce leakages in the water resources systems for optimal usage and minimum wastage. The various components and techniques dealing with smart water management have been discussed.

Chapter 10 - Solar Energy is the most readily available and practically inexhaustible power source that has helped in replacing conventional fossil fuels. It has started playing a major role in the sustainable development of the nation along with eliminating the adverse impacts on the environment. Because of its numerous advantages, most of the developed nations have switched over to it on a magnanimous scale. Following this trend, India has also done its due. Because of the geographical location of the country, solar energy has huge potential for development in this region, and many hotspots have been identified where this energy can be tapped efficiently. The chapter reviews the scope, potential, and steps taken to date by the government in this sector especially the Jawaharlal Nehru National Solar Mission, which has proved to be a guiding light in the path of sustainable development.

Chapter 11 - Bangladesh is facing a water crisis that has put economic development, livelihoods, human well-being, and ecological sustainability at risk. In recent times, this country has witnessed rapid, unplanned urban expansion accompanied by serious environmental degradation. Conventional urban aquatic bodies are the ponds, canals, and any types of water bodies available in an urban city. The authors attempt to illustrate and examine the management of urban water bodies in the commercial capital city of Bangladesh by focusing on their characteristics and spatial distribution, using patterns and perceived health impacts on users and causes of disappearance and consequences of losses and degradation. The number of water bodies in the commercial capital city was found declining gradually. The author found that the proper coordination of urban autonomous bodies, citizen awareness, community movement, the inclusion of urban management-related laws, rules, and regulations, and legislative strengthening would be reducing the negative impacts of unplanned urbanization in urban areas of Bangladesh. The authors have recommended effective implementation of the existing laws and regulations at the field level for a sustainable and healthy environment for the inhabitants.

Part I: Socio-Political and Religious Trends (SPRT)

Chapter 1

The Saints of Persia

Theodore J. Drizis, MD, PhD*
Independent Researcher, Kalamata, Greece

Abstract

The aim of this word is to present the eponymous and anonymous Saints of Persia as well as a historic religious analysis of the situation in it, mainly in 4th c. CE, bringing the necessary witnesses as target. The material of this work are monthly books of the Saints of Christian Church, calendars, Holy Bible, ancient Greek literature, other historic and religious writings, encyclopedias, journals and electronic pages of the world wide web. Method is the critical analysis of text. Results are the presentation of the Saints by month, the Christians' persecutions in Persia, the presentation of Zoroastrianism and Zoroaster, the adoration of fire, the position of fire in Judaism and Christianity, the reference on account of Magoi – Magians and their martial against Christianity and the short-lived its recognition by Persian empire but in its heretical of Nestorianism form. As a conclusion, refers to the inflicted damage of a country from the persecutions of the State on a healthy religious reform.

Keywords: saints, Persia, Christianity, Zoroastrianism, Magoi, Magians

Introduction

Persia always was opposed to Christianity and to Hellenism perhaps due to ethnicity or to some other cause. Nonetheless, there were Persians, who

* Corresponding Author's Email: thdrizis@ath.forthnet.gr.

In: Handbook of Uncertainty in Eurasian Forecasting (HEF)
Editors: Saeid Eslamian, Mir Bintul Huda and Faezeh Eslamian
ISBN: 978-1-68507-966-6

preserving their national origin, lived as Christians, proclaimed their Christian property, and remained steady in their relations with Jesus Christ until their death, becoming Christian martyrs, while the official Persia always supported the paganism, the wise men – Magi, the Zoroastrianism (Mazdayasna) and the Mohammetism.

Persians despite the expulsion of paganism and Zoroastrianism from themselves, in majority, they did not accept Christianity which it did not impose to them its acceptance but it calls all people to embrace it voluntarily and without pressure, maintaining their local characteristics as language, culture and nationality. Nevertheless, they adopted the Mohammetism due to the force and pressure of it on them.

Persians also, in major part, rejected any Christian and Greek element. They did not develop philosophy and critical thought but they insisted on myths, idolatry, magic, pietism, temporary gain and on the fear of the Mohammedians' sword. Furthermore, Persians wanted to be different from Christians due to their arrogance, to national causes as the fear of the deprival of their language and customs, as well to their reaction to the superiority of Christianity.

The Saints of Persia in Calendar

In Persia there was Christian Community. Before Christ there were Kings who recognize the true Jews' God, as 1) Nebuchadnezzar (Daniel, 3. 28 – 33; 4. 31 – 34), 2) Belshazzar, Nebuchadnezzar's son (Daniel 5. 29 – 30), 3) Darius I, (Daniel, 6. 25 – 28), and 4) the prophet Balaam (Numbers, 23. 7 – 8; 24. 17 – 18).

In Christ's Era, Persians Christians were known and memorable as Saints – Martyrs.

Per month they are:

January

Jacob (Bishop of Nisibis) 13/1, Anastasius the Persian 22/1, Aphraates the righteous 29/1.

February

Abramius (Bishop Arbil and Sophines of Persia) 4/2, Maruthas the righteous and Theodosius' I ambassador in Persia 16/2; the Saints – Martyrs in Persia whose the relics St. Maruthas gathered and in their name established the town Martyropolis (Miepherkin - Silvan) on the boundaries of Turkey - Persia – Armenia. An anonymous woman of Persia 19/2. Sadok bishop and another 128 Persian martyrs with him 20/2.

March

Saint Martyrs Christin from Persia 13/3. S. Martyrs Codratus, Theodosius, Immanuel and another 40 from the East (26/3). The nine S. Martyrs in Persia Zanitha, Lazar, Masrutha, Narsi, Elijah, Mares, Avivus, Sebeyth and Sabbas in the city of Marmiabah during the Sabor's era, King of Persia (27/3). The S. Martyrs Jona and Barah in Marmiabah too (27/3). The S. Martyrs Auda bishop, Benjamin deacon and many others, Ormisd of Achaemenids' dynasty, between them, who all of them were eaten by rats and cats in prison (31/3).

April

The S. Martyr and Bishop Simeon, his sister Pherbuth with her maidservant, another his sister, and another one thousand Martyrs during the King Sabor's reign (4/4). The 120 S. Martyrs in Persia (6/4).

The 295 S. Martyrs dead on seizure in Persia during Sabor's reign due to their refusal to worship the sun and the fire (9/4).

The S. Martyrs Badem, archimandrite of a monastery with his seven disciples (9/4).

The 400 S. Martyrs, Desan Bishop, Avjesus (9/4).

The S. Martyrs Jacob priest, Aza deacon during Sabor's reign due to refused to adore the sun and the sky (10/4).

The S. Martyr Elephtherius the Persian (13/4).

The S. Martyrs Simeon priest, Abdul priest, Gothazat (of the king's palace), Phusic (of the king's palace) and another one thousand one hundred fifty S. Martyrs during Sabor's reign (17/4).

May

The S. Martyr Bada (clergyman) the Persian (1/5).

The S. Great Martyr Irene from Persia, called before Penelope. Due to her became Christians her father and King Licinius, her mother Licinia, Sapor, the Governor of the province named Baud and 146.000 Persians according the tradition (5/5).

The S. Apostle Simon, the Zealot, proclaimed the Christian faith in Persia, too (10/5).

The S. Martyr Abdisus Bishop together with another 38 Persian Martyrs and S. Abdas (16/5).

The S. Kallicus Mag (24/5).

June

The S. Martyrs Thecla, Mariamn, Martha, Maria, Ennatha in Persia (9/6).

The S. Martyrs Sabel, Manuel, Ismael, Persians in gender, were massacred by sword from the Byzantine Emperor Julian in 362CE (17/6).

July

The S. Martyrs Golinduh from Persia living with a Senior Mag (13/7).

The S. Martyr John, Persians' King, in 836 CE with another three Arab soldiers Martyrs were killed by Persian Mohammedians after the denial of Mohammetism by them (19/7).

The S. Theodosius the Young, who's Commissar in his youth was the Persian Emperor Yazd(e)gerd I, contributed to the development of Christianity in Persia by Antiochus whom the Persian Emperor Yazderd I sent at Constantinople for learning and establishment of Christianity in Persia (29/7).

August

The S. Martyr Ias and the nine thousands S. Martyr with her, during King Sapor's reign, being prisoners by the chiefs of Magicians (4/8).

The S. Martyr Eudocia (4/8).

The S. Rightous and Martyr Dometius the Persian (7/8).

The S. Dosas or Sodas or Dose (20/8).

September

The S. Martyr bishop Abdu was killed from a chief of Magicians (5/9).

The S. Martyr Godelia, the S. Martyr Gobdelaas, son of the King of Persians Sapor, the S. Martyr Casdoa, daughter of the King of Persians Sapor, the S. Martyr Dadas, related of the King of Persians Sapor, the S. Martyr Gargal, exmagician, and the S. Martyr Casdoos or Casdias (29/9).

October

The Persian S. Martyr Mamelchth, expriest of paganic Greek temple she was stoned by Greeks. She had a sister, Christian too (5/10).

The S. Martyr deacon Benjamin (13/10).

The S. Martyr Sadoth and another hundred twenty S. Martyrs together with him, while one thousand seventy people in Persia accepted Christianity too (19/10).

November

The S. Martyrs John Bishop, Jacob Priest (1/11).

The S. Martyrs Acindynos, Pegasios, Aphthonios, Elpidophoros, Anempodistos, the King Sapor's mother and another seven thousands S. Martyrs who were beheaded ((2/11).

The S. Martyrs Achepsimas, Joseph, Aithalas (3/11).

The S. Martyrs Achemenides, the Confessor, during the Yazderd's reign, Gororane's son King of Persian (3/11).

The S. Martyr Milos (or Miles or Milisios) – Bishop, two S. Martyrs his disciples, the S. Martyr Gennadios clergyman and Confessor and S. Martyrs Bishops of Synod in Ctesiphon (Tisfun) 10/11.

The S. Martyr Nirsa (Bishop). The S. Martyr Joseph, Bishop and his disciple and another with him in Persia. The S. Martyrs Hisac, John and Sapor were stoned. The S. Martyrs Geythazed and another three were bayoneted. The S. Martyrs, three virgins Thecla, Bauth and Denah were killed by a sword.

Together with them Bishops and Priests as well as virgin nuns all from a monastic order were killed by various afflictions and torments (20/11).

The Rightous Silos Bishop in Persia (26/11).

The S. Martyrs Jacob the Persian, his wife and his mother (27/11).

The S. Martyr John, clergyman in Persia (29/11).

December

The S. Martyr Ananias the Persian (1/12).

The S. Narsis the Persian (9/12).

The S. Martyrs Aithalas, Achepseis deacon from Arbil of Persia (11/12).

The S. Martyr Barsaba beheated in Persia (11/12).

Discussion

Discussing the above list of the Saints of Persia, who had been died due to their Jesus Christ's acceptance and confession under various tortures, we can see they were from both gender, men and women. Also, some of them had been girls who lived the devoted to Jesus Christ virginal life. The period for the referred Saints is the 4th c. CE, except of four Saints who lived in 9th c. CE and were killed in 836 CE, particularly the Christian Persian King John, and three Christian Arab soldiers, fact that means there were Christians between 4th – 8th c. CE without be excluded in next years. The Saints were of all layers, clergymen, laymen, a Magian, a King, members of a Royal family. We can see existence of Monastery and massacres en masse.

The Persians' religion initially, was the adoration of the fire, of the sun and of the stars, adoration which was incorporated to Zoroastrianism and were managers the Magians (Magoi) as we can see from the aforementioned Saints' calendar.

It is known, in New Testament is referred the Magians' (wise men) worship to Jesus Christ travelling from the East to Judea (Bethlehem) in spite of all the difficulties and the fear of the march, of the weather, of the existence of bandits and robbers, in order to know and to honor Him as King of Jews, due to they saw his star in east with march to west (Holy Bible, New Testament, Matthew, 2. 1 – 2). The fact that the Magians had knowledge of the star of the King of Jews, it was neither unexpected (causal) nor an inspiration of that moment. They knew the history of area and the era of that region, history that was saved in oral tradition or in writings as in Old Testament where it says: "...a star shall come out of Jacob, and a scepter shall rise out of Israel; it shall crush the borderlands of Moab, and the territory of all the Shethites..." (Holy Bible, Old Testament, Numbers, 24.17). This witness is a Balaam's prophecy or oracle. Balaam came from Mesopotamia (Numbers, 22.5) near Persia, he was a prophet dated about 15th – 12th c. BC in relation to exodus of the people of Israel from Egypt. This prophetical witness was conserved by the Magians – wise men – of that area and era and was verified and confirmed during the Jesus Christ birth in Bethlehem (New Testament, Matthew, 2. 1 – 12). This verification and confirmation of Balaam's prophecy would be transmitted by these Magians through generations and would be known by the Jesus Christ's disciples and the Gospels in order to be created the substratum for the acceptance of Christianity without any oppression, violence and war. However, the Magoi and the majority of Persians rejected the peaceful Christianity while accepted without resistance the oppressive Mohammetism of Arabs[1].

To the rejection and the persecution of Christianity in Persia a main cause is the political, despite some intervals of tolerance and recognition by the Kings, let it be with its heretical form, for instance Nestorianism. In an article, university Anglo-Saxon professors mention[2]: "...The Roman Empire became Christian and the Syrochristian populations of Mesopotamia and Babylonia started to feel sympathy to roman policy due to religious causes... Consequently, Sassanids Emperors perceived the necessity to secure the Zoroastrianism and endeavor to enforce it. Every religion, heresy and sect was persecuted from the state, the inclination from the official faith was considered crime and afterwards persecutions started against contraries, especially against Christians. Thus, the competition between Persia and Byzantium took religious dimension, too. The war was inevitable. It started whereupon Shapur II in 337 CE...After some decades Yazdegerd I took the chair of State. His intial tendency to reveal tolerance in Christianity and Judaism, provoked the Gentiles' reaction. Due to Gentiles' revolt and to increasing Christians' fanaticism, Yazdegerd (I) was constrained move on repressive policies. ...Bachram...failing on the war against Byzantium (421 – 422 CE), associated himself with Byzantium a centenarian peace and guaranteed the religious

---

1 *New Encyclopedia, v. 20,* p. 362. Editions "Malliaris – Pedia," Thessaloniki 2006.

2 *Encyclopedia, Papyros – Larousse – Britannica,* lemma "Persia," v. 42, pp. 265 – 266. Edition "Papyros," Athens 2007.

Christians' freedom. Nestorianism had already between Christians of Persia and it was recognized as the official form of Christianity in the Sassanids' Empire…long-drawn and exhaustive operations decreased efficiently the armed forces of Persia and Byzantium. Thus, the way remained open and free for a new army, the Mohammedians Arabs, who hit both Empires, nevertheless their religions too. The battle at Nahavand (642 CE) in South of Hamedan extinguish any their expectation…The fall of Empire defined the future of the religion. Mohammedians officially showed tolerance to Zoroastrianism although persecutions did not disappear. However, this religion, gradually was disappeared except some its followers. …"[3]

In my opinion, the Mohammedians Arabs' expedition did not have religious only character but political too, since Arabia, little time before, was a part of the Persian Empire.

Daniel, Bishop in Kaisariani – Byronas – Ymettus (Attica), in his report on account of Christianity in Persia, says: "…During the 3rd c. CE there was a grand increase of the Christianity in Persia…Surely, the offended clergy of Zoriastrianism opposed against Christianity. Nonetheless, as Christianity was persecuted in Roman Empire, Persian State did not display specific hostility to it. The policy to Christianity was changed when it dominated in Roman Empire. The adjacent to Persia populations, as Armenians and Syrians, due to had antipersian spirit, often received attacks from Persia. Besides, due to these adjacent populations were Christians, in Persia started to develop an antichristian spirit. This change had serious consequences to Christians of Persia. In 339 CE King Shapur II started a terrible persecution, which lasted forty years, until 379 CE and it was repeated in 420 CE, during the ruling of King Yazdegerd. Shapur II was the unique King, who imposed by violence Zoroastrianism in Persia executing the followers of the other religions, particularly the Christians, in the era when Christianity became the official religion of the Roman Empire by the Emperor Constantine I. The causes probably were political, too, since Shapur II considered the Roman Empire as the greatest danger for Persia. Among the Persian Saint Martyrs during this persecution were members of the old and former Achaemenids' royal family. After the Persians' repellence by Greek – Roman Heraclitus' army from the

---

[3] Adrian David Hugh BIVAR, Profess. of Iranic Studies in University of London (Britannica). Mark J. Dresden; Emer. Prof. of Iranic Studies in University of Pensylbania, editor of "Modern Persian Reader" (Britannica); Seton H.F. Lloyd, Emer. Prof. of Western Asian Archaeology, of University of London (Britannica); T. Cuyler YOUNG Jr. Prof. of Studies of Middle East, of University of Toronto (Britannica). In Encyclopedia, *Papyros – Larousse – Britannica*, v. 42, pp. 265 – 266. Edition "Papyros," Athens 2007.

sacked regions of the Greek – Roman Empire in the beginning of 7th c. CE, Christianity was recognized in Persia as religional minority and by this manner, the Persians Christians lived more quietly. This condition was facilitated by the fact that the Christians of Persia after the fourth Ecumenical Synod, 481 CE, were incorporated to Nestorianism, a heresy of Christianity, which was abolished in Byzantine Empire and consequently without any suspicion of rebellion to Persian regime. Thus, Christianity was transmitted in all over the regions of Persia and the Christians were above of one million. Certainly, the Christians' destiny in Persia was depended on the relations between Persian and Byzantine Empire. Any straining between them provoked Christians' persecutions by Persian State due to incitement of Zoroastrian clergy but the presence of Christianity in Persia was continuous and stable until its conquest by Mohammedians."[4]

Concerning the religion in Persia, before Christianity and Mohammetism, was the Zoroastrianism, established in 7th – 6th c. BC, by Zoroaster. This religion kept the pre-existed worship of the fire and of the sun[5].

The religious and philosophic interpretation and importance, which had the fire and the sun, did not be exclusiveness only of the Persians in ancient era, as it is obvious from the search and the study of the history.

The adoration of the fire was reflected on "...the name of deities as Hestia and Hephaestus in ancient Greece. At ancient Rome was the custom of Hestiads...Something proportionate believed some races of the Indians of America. ..."[6] The fire is mentioned to Biblical texts too, from 15th c. BC by Moses[7]. For instance, "...For the Lord your God is a devouring fire, a jealous God..." (Holy Bible, Deuteronomy, 4. 24), as well in other part Moses again says "...the Lord your God is the one Who crosses over before you as a devouring fire; he will defeat them and subdue them before you, so that you may dispossess and destroy them quickly,..." (Holy Bible, Deuteronomy, 9. 3)[8].

Concerning the concept of the sun with its natural and philosophic significance, it is present to the Israelites, too. With its natural meaning it is

---

[4] Bishop, Daniel, *Persecutions against the Church*, pp. 93 – 95. Edition "Apostoliki Diaconia," Athens 2018.

[5] Encyclopedia, *Papyros – Larousse – Britannica*, v. 22, pp. 121 – 122. Edition "Papyros." Athens 2007.

[6] *New Encyclopedia, v. 22*, pp. 112 – 113. Edition "Malliaris pedia." Thessaloniki 2006.

[7] G. P. Sotiriou, *Pedagogue in Christ – Introduction to Old Testament*. 2nd Edition, p. 12. Editor "Apostoliki Diaconia." Athens 2005.

[8] Holy Bible, *Deuteronomy,* New Revised Standard Version, pp. 160 and 165. American Bible Society. New York 1991.

presented to Moses (about 1500 BC) in the phrase: "…And when you look up to the heavens and see the sun, the moon, and the stars, all the host of heaven, do not be led astray and bow down to them and serve them, things that the Lord your God has allotted to all the peoples everywhere under heaven…" (Holy Bible, Deuteronomy, 4. 19)[9]. With its philosophic and Spiritual significance is present in Prophet Malachi (474 BC)[10] with the phrase: "…See, the day is coming, burning like an oven, when all the arrogant and all evildoers will be stubble; the day that comes shall burn them up, says the Lord of hosts, so that it will leave them neither root nor branch. But for you who revere my name the sun of righteousness shall rise, with healing in its wings. You shall go out leaping like calves from the stall…" (Holy Bible, Malachi, 4. 1 – 2; or Masoretic text 3. 19 – 20)[11]. Besides this, the Prophet Isaiah (770 – 700 BC)[12] says: "…Arise, shine; for your light has come, and the glory of the Lord has risen upon you…The sun shall no longer be your light by day, nor for brightness shall the moon give light to you by night; but the Lord will be your everlasting light, and your God will be your glory. Your sun shall no more go down, or your moon withdraw itself; for the Lord will be your everlasting light, and your days of mourning shall be ended…" (Isaiah, 60. 1; 60. 19 – 20)[13].

On account to Egyptians, Diogenes Laertius (3rd c. CE)[14] says: "…The philosophy of the Egyptians is described as follows…The sun and the moon are gods bearing the names of Osiris and Isis respectively…" (Lives of eminent philosophers, I. 10 Prologue)[15].

Respecting the Greek philosophy on the fire and the sun, firstly Diogenes Laertius saved an Heraclitus' quotation: "…All things are composed of fire, and into fire they are again resolved… (Lives of eminent philosophers, Heraclitus, IX. 9)[16]. Secondly, the writer or the writers under the name S.

---

[9] Ibidem, p. 160.

[10] G. P. Sotiriou, *Pedagogue in Christ – Introduction to Old Testament*, p. 143. 2nd Edition, Editor "Apostoliki Diaconia." Athens 2005.

[11] *Holy Bible,* Malacchi, New Revised Standard Version, p. 878. American Bible Society. New York 1991.

[12] G. P. Sotiriou, *Pedagogue in Christ – Introduction to Old Testament*, pp. 123 – 124. 2nd Edition, Editor "Apostoliki Diaconia." Athens 2005.

[13] *Holy Bible, New Revised Standard Version*. Isaiah, p. 679 – 680. American Bible Society. New York 1991.

[14] P., Kroh. *Lexikon der Antiken Autoren [Lexicon of Ancient Authors]*, translated in Greek by D. Lipourlis – L. Tromaras, p.143. University Studio Press. Thessaloniki 1996.

[15] Diogenes Laertius, *Lives of Eminent Philosophers*, translated by R. D. Hicks, v. 1, pp. 12 – 13. Harvard University Press, Cambridge (MA) – London (EN), 1972. Loeb Classical Library.

[16] Ibidem, v.2, p. 415.

Dionysius Areopagite[17] (1st – 5th c. CE) mention(s): "…according to the old word, the sun being god and creator of the universe is a guardian of the visible world..." (On Divine Names, 4. 4)[18].

Likewise, in Ancient Greek Literature again, there are references to the Magi (Magians) and Zoroaster. Diogenes Laertius says: "There are some who say that the study of philosophy had its beginning among the barbarians. They urge that the Persians have had their Magi, the Babylonians or Assyrians their Chaldaeans, and the Indians their Gymnosophists; and among the Celts and Gauls there are the people called Druids or Holy Ones…The date of the Magians, beginning whith Zoroaster the Persian...But the advocates of the theory that philosophy took its rise among the barbarians go on to explain the different forms it assumed in different countries. As to the … Gymnosophists and Druids we are told that they uttered their philosophy in riddles, bidding men to reverence the gods, to abstain from wrong-doing, and to practice courage…the Gymnosophists at all events despise even death…the Chaldaeans apply themselves to astronomy and forecasting the future; while the Magi spend their time in the worship of the gods, in sacrifices and in prayers, implying that none but themselves have the ear of the gods. They propound their views concerning the being and origin of the gods, whom they hold to be fire, earth, and water; they condemn the use of totems (comment: mine translation of this word), and especially the error of attributing to the divinities difference of sex…With the art of magic they were wholly unacquainted, according to Aristotle in his *Magicus* and Dinon in the fifth book of his *History*. Dinon tell us that the name Zoroaster, literally interpreted, means "star-worshipper" (At this point, the remarkable translator R. D. Hicks, a Fellow of Trinity College in Cambridge, denotes: This popular etymology, though wide-spread, is erroneous, the true form of the prophet's name being Zarathustra, almost certainly derived from *zarath* = old [Zend stem, parallel to γεροντ-] and *ustra*=camel)[19]…Aristotle in the first book of his dialogue *On Philosophy* declares that the Magi are more ancient than the Egyptians; and further, that they believe in two principles, the good spirit and the evil spirit, the one called Zeus or Oromasdes, the other Hades or

---

[17] S. Dionysius Areopagite, *On Divine Names*, translation in Modern Greek by Ign. Sakalis, introduction by Mar. Nikolaou, pp. 7 – 11. Edition "Papazisis," Athens 1978.
S. Dionysii Areopagitae, *De Divinis Nominibus [On Divine Names]*, cap. IV, § IV, col. 700. Patrologiae, Series Grecae, Accurante J. P. Migne. Parisiis 1858.

[18] Ibidem, pp. 58 – 59, edition "Papazisis."

[19] Diogenes Laertius, *Lives of Eminent Philosophers*, v. 1, pp. 8 – 9. Harvard University Press. Cambridge (MA) - London (EN), 1972. Loeb Classical Library.

Arimanius…Theopompus in the eighth book of his Philippica…says that according to the Magi, men will live in a future life and be immortal, and that the world will endure through their invocations…But Hecateus relates that according to them the gods are subject to birth…in the creed of the Magi…statues of the gods should be destroyed by Xerxes was natural enough…" (Diogenes Laertius, Lives of Eminent Philosophers, I. 1 – 9)[20].

On account of the Zoroaster's time of life there is disagreement and there are various versions, ancient and newer. Diogenes Laertius (3rd c. CE) gives Hermodorus' (4th c. BC) report that Zoroaster lived 5000 years before the fall of Troy, as well as Xanthus' (5th c. BC) that Zoroaster lived 6000 years before Xerxes' expedition[21]. Another researcher states the opinion, Zoroaster lived during 7th – 6th c. BC, opinion based in historical facts and accepted more and more[22]. On account to this, Bruce Lincoln notes: "…At present, the weighty opinion between the researchers is inclined to the end of the second millennium or at the beginning of the first millennium BC, despite those who considerate a date in the 7th c. BC. …"[23]

Diogenes Laertius presents opinion of elder of him authors, respecting the Jews' origin that the Jews' lineage is from Magi. I think it is in error, due to, firstly, Moses lived in 1500 – 1200 BC about[24], and the Zoroaster, the first of Magi, lived in 10th - 7th c. BC. Secondly, Magi believe, concerning their theology, according Aristotle and referring by Diogenes Laertius, in two principles, the good spirit and the evil spirit, while the Jews believe to one God only (Bible, Old Testament, Exodus, 20. 2; Old Testament, Deuteronomy, 5. 6; Deuteronomy 6. 4; Job 1. 6)[25].

Other reports from ancient Greek writers on account of Zoroaster and Magi are, firstly, Aeneas Gazaeus (5th c. CE)[26] in his work "Theophrastus," is referred that "Zoroaster predicts will be time in which will become

---

[20] Ibidem, pp. 2 – 11.

[21] Ibidem p. 5.

[22] Shahbazi, Shapur A., *The Traditional Date of Zoroaster Explained, Bulletin of the School of Oriental and African Studies*, 40 (1): 25 – 35, 1977.

[23] Lincoln, Bruce, *Death, War, and Sacrifice, pp.149 – 150. In Studies in Ideology and Practice*. Chigago University Press, 1991.

[24] Sotiriou, G.P. *Pedagogue in Christ – Introduction to Old Testament*, p. 12. 2nd Edition "Apostoliki-Diaconia," Athens 2005.
Kalantzakis, S.E. *Introduction to Old Testament*, p. 26. Edition "Pournara," Thessaloniki 1985.

[25] *Holy Bible, New Revised Standard Version*, pp. 65, 161, 162, 446. American Bible Society. New York 1991.

[26] Kroh, Paul. *Lexikon der Antiken Autoren [Lexicon of Ancient Authors]*, translated in Greek by D. Lipourlis – L. Tromaras, p. 37. University Studio Press. Thessaloniki 1996.

resurrection all of dead, saying that Theopompus knows and he teach it at the others"[27]. To keep in mind for this saying, Zoroaster probably was influenced by Jew King David, 10th c. BC[28] (Psalm 16. 8 – 11; psalm 86. 13)[29], Isaiah, 8th c. BC[30] (Isaiah, 14. 9; 26. 19; 28. 18; 53. 10 – 12)[31] and by Ezekiel 6th c. BC[32] (Ezekiel, 37. 1 – 14)[33]. Secondly, Plutarchus (1st – 2nd c. CE)[34] in his work "Isis and Osiris" notes: "...Theopompus says that, according to the sages (magians in Greek text), one god is to overpower, and the other to be overpowered, each in turn for the space of three thousand years, and afterward for another three thousand years they shall fight and war, and the one shall undo the works of the other, and finally Hades shall pass away; then shall the people be happy, and neither shall they need to have food nor shall they cast any shadow...such, then, is the character of the mythology of the sages (magians in Greek text)...(Plutarch, moralia, 370. B – D)[35]. In this second case, we can see the cycle of three thousand years which gave an error for the Zoroaster's age, as well the magians' conviction, for anoher, beyond death, life with a probable Jewish influence, as it has been said above on Aeneas' Gazeus report. Thirdly, Herodotus (5th c. BCE)[36] in his work "Persian Wars" notes the Magians' kingship in Persian Empire and their interference and

---

[27] Gazaei, Aeneae. *Theofrastus*, Patrologiae – series Graeca, v. 85, cap. 662A, column. 996. Accurante J. – P. Migne. Parisiis 1860. Athens 2003.

[28] Kaimakis, D. *Short Introduction in Old Testament – Selection of Units from the Bible of Jerusalem*, p. 53. Uniprint Hellas. Thessaloniki 1991.

[29] *Holy Bible*, *New Revised Standard Version*, pp. 489, 536. American Bible Society. New York 1991.

[30] Kaimakis, D. *Short Introduction in Old Testament – Selection of Units from the Bible of Jerusalem*, p. 149. Uniprint Hellas. Thessaloniki 1991.

[31] *Holy Bible*, *New Revised Standard Version* pp. 633, 643, 645, 673. *Old Testament* in Greek, pp. 709, 716, 717, 733. Editions "Zoi," Athens 1986. American Bible Society. New York 1991. *Holy Bible*, New Revised Standard Version, p. 794. American Bible Society. New York 1991.

[32] Kaimakis, D. *Short Introduction in Old Testament – Selection of Units from the Bible of Jerusalem*, p. 163. Uniprint Hellas.Thessaloniki 1991. Paul Kroh, *Lexicon der Antiken Autoren*, translated in Greek by D. Lipurlis – L. Tromaras, p. 385. University Studio Press. Thessaloniki 1996.

[33] Plutarch, *Moralia*, translated by Frank Cole Babbit, v. V, pp. 114 – 117. Harvard University Press. Cambridge (MA) – London (EN) 1936. Loeb Classical Library.

[34] Kroh, Paul. *Lexicon der Antiken Autoren [Lexicon of Ancient Authors]*, translated in Greek by D. Lipurlis – L. Tromaras, p. 211. University Studio Press. Thessaloniki 1996.

[35] Herodotus, *Persian Wars*, translated in English by A. D. Godley, v. ii, pp. 154 - 155, 184 – 187. Harvard University Press. Cambridge (MA) – London (EN), 2006. Loeb Classical Library.

[36] Kroh, Paul. *Lexicon der Antiken Autoren [Lexicon of Ancient Authors]*, translated in Greek by D. Lipurlis – L. Tromaras, p. 434 – 435. University Studio Press. Thessaloniki 1996.

participation in politics (Herodotus, Persian Wars, III. 126, 150, 153)[37]. Forthly, Strabo (64 BC – 25 CE) in his work "Geography" considers the Magi of Persia like the Jewish Prophets (Strabo, Geography, XVI, 2. 39 or C 762)[38].

Thus, in Zoroastrianism, in my opinion, there are rudiments of adoration of fire and sun as well there are rudiments of Judaism not only from Zoroaster, who was the first of the Magi, but from his followers too. The same occurs with Mohammetism, which has elements from Judaism, Paganism, respecting for instance of Kaaba, according to various writers[39], and Christianism. Besides, the Zoroaster's theory is deprived of historical elements but in contrary it is filled of mythicized and mythological elements[40]. Consequently, Zoroastrianism was not a religion clearly Persian.

An occasion for peace and cooperation between Byzantine and Persian Empire and for the establishment of Christianity in Persia coexisting among other religions has been perished, due to slaughter of the Christian ambassadors of Persia, Manuel, Savel and Ismael by paganist Byzantine Emperor Jiulanus in 17 June, 362 CE, for their refusal to sacrifice to idols[41].

Recently, by the revolution of Mohammedian clergy in Persia during 1978 – 1979 CE, the Christians of Persia were the goal of organized enmity and violence. Many Christians of Persia abandoned it, while those who remained they suffered, especially those of South Persia. Christian Martyrs of this time are considered the Avastu Sayah, a clergyman of Anglican Christian Church at Sira, who was assassinated in his office in 19 February, 1979 CE[42], as well as Bahram Tafti 24 years old, son of the Anglican Bishop Dehani Tafti who was assassinated on the roads of Teheran in 6 May, 1980 CE[43], while his father and his mother had been an attempt of their assassination by Mohammedians in October 1979.

---

[37] Strabo, *Geography*, translated in English by Horace Leonard Jones, v. VII, pp. 288 – 289. London – William Heineman, New York – C. P. Putnam's Sons. MCMXXX.

[38] Strabo, *The Geography*, with an English translation by Horace Leonard Jones, v. VII, p. 289. Loeb Classical Library (LCL 241). London: William Heinemann CTD. New York: G. P. Putnam's Sons. MCMXXX.

[39] Wikipedia, *Kaaba, History – Other views of origine – Pre-Islamic era, References* 39 – 62. En.wikipedia.org/wiki/Kaaba.
New Encyclopedia, v. 15, p.309. Edited by "Malliaris – Pedia," Thessaloniki 2006; *Encyclopedia "Papyros – Larousse – Britannica,"* v. 35, p.386. Edited by "Papyros," Athens 2007.

[40] *Encyclopedia "Domi," lemma "Avesta,"* v. 1, p. 50. Editions "Domi," Athens 2002 – 2004; *Encyclopedia "Papyros – Larousse – Britannica," lemma "Avesta,"* v. 1, p. 102, lemma "Zoroaster" and "Zoroastrianism," v. 22, pp. 121 – 130. Editions "Papyros," Athens 2007.

[41] *Daily Readings for Month*, Jun 17, 2019. Greek Orthodox Archidiocese of America.

[42] http://articleeighteen.com/features/2907/.

[43] http://articleeighteen.com/features/6011/; http//: iranicaonline.org/artricles/christianity-viii/.

Conclusion

Taking all the above in consideration, I am convinced, that religions that are not harmful to the citizen's country, must not be prohibited and their members must not be prosecuted. Otherwise, the persecution undermines the foundation of the state, deprives human personnel of it and renders the country defenseless to a foreign enemy, the fact that was verified to Persia with the Mohammedians from Arabia in 642 CE[44].

References

Basiliadis, N. P., *Archaeology and Bible*, edited by "Sotir," Athens 1999.

Bishop, Daniel, Metropolitan Caisarianis – Byronus – Ymettus, *Persecutions against the Church*, edited by "Apostoliki – Diaconia." Athens 2018.

Boyce, M., *A History of Zoroastrianism*. Brill 1996.

Calendar 2008, Saint George's Holy Cell, Holy Mountain Athos – Karies 2007.

Diogenes Laertius, *Lives of Eminent Philosophers*, translated by R. D. Hicks. Harvard University Press, Cambridge (MA) – London (EN), 1972. Loeb Classical Library.

Dionysii Areopagitae, S., *De divinis nominibus [On Divine Names]*, in Patrologia – Series Grecae, v. iii, accurante J. – P. Migne. Parisiis 1858. Reprint Athens 2008.

Encyclopedia "Domi." Editions "Domi," Athens.

Encyclopaedia Iranica. Encyclopaedia Iranica Foundation, New York.

Encyclopedia "Papyros – Larousse – Britannica." Edition "Papyros." Athens 2007.

Eortologio.com, www.eortologio.gr

Gazaei, Aenaei, *Theophrastus,* Patrologiae – Series Graeca. Accurante J. P. Migne, Parisii 1860. Reprint 2003.

Greek Orthodox Archdiocese of America, www.internet.goarch.org.

Herodotus, *The Persian Wars*, translated in English by A. D. Godley, v. ii. Harvard University Press, Cambridge (MA) – London (EN), 2006. Loeb Classical Library.

Holy Bible, New Revised Standard Version, American Bible Society, New York 1991.

Iranica Online, *CHRISTIANITY viii. Christian Missions in Persia*. www.iranicaonline.org/articles/christianity-viii.

Kaimakis, D. *Brief Introduction to Old Testament – Selected Units of the Bible of Jerusalem*, translated by Thessaloniki 1991.

Kalantzakis, Stavros E., *Introduction to Old Testament* – General Introduction. Editions "P. Pournaras." Thessaloniki 1985.

Karavidopoulos, John D. *Introduction to New Testament*. Edited by "P. Pournara." Thessaloniki 1991.

Kroh, Paul. *Lexicon der Antiken Autorem*, translated in Greek by D. Lipourlis – L. Tromaras. University Studio Press, Thessaloniki 1996.

---

[44] Encyclopedia *Papyros – Larousse – Britannica*, v. 41, p. 266. Edition "Papyros," Athens 2007.

Lincoln, Bruce. *Death – War – Sacrifice. Studies in Ideology and Practice*. University of Chicago 1991.

Meniaion, 2nd edition. Ecclesiastic Library "Fos." Athens.

Meniaion, edited by Apostolic Diaconia of the Church of Greece. Athens 1992.

New Encyclopedia. Editions "Malliaris – Pedia," Thessaloniki 2006.

Nestle, Aland. *Greek – English New Testament*, 9th revised edition, including papyri 98 – 116. Deutsche Bibelgesellschaft, Stuttgart 2001.

Nicodemus, Athonites of Saint Mountain, *Synaxaristes (Gathering of Saints) of Twelve Months Of The Year*. Edition "Orthodox Kypsely," Thessaloniki 2003.

Orologion Mega, editions "Astir" A & E. Papadimitriou. Athens 1973.

Plutarchus, *Moralia – Isis and Osiris*, vol. v, translated in English by Franc Cole Babbit. Harvard University Press, Cambridge (MA) – London (EN), 1936. Loeb Classical Library.

Pyne, Grenfell Bernard et Arturus Hunt, *Hellenica Oxyrhynchia cum Theopompi et Cratippi Fragmentis*. Oxonii e Typographeo Clarendoniano. MDCCCCIX.

Saints' Calendar (Agiologion) of Year 2011. Editor "Saitis." Athens 2011.

Shahbazi, Shapur A., The Traditional Date of Zoroaster Explained, Bulletin of the School of Oriental and African Studies, 40 (1): 25 – 35. 1977

Sotiriou, G. P., *Pedagogue at Christ – Introduction to Old Testament*. 2nd Edition. Editor "Apostoliki – Diaconia." Athens 2005.

Sotiriou, G. P., *The word of Grace – Introduction to New Testament*. Editor "Apostoliki – Diakonia. Athens 1992.

Stausberg M., Yuhan Sohrab – Dinshaw V., Tessman A., *The Wiley – Blackwell Companion to Zoroastrism*. John Wiley and Sons, 2015.

Strabo, *Geography,* translated in English by Horace Leonard Jones, v. vii. Edited by WilliamHeineman (London) – C. P. Putnam's Sons (New York) 1931, now by Harvard University Press Cambridge (MA) – London (EN) 2005. Loeb Classical Library.

Symeonis Metaphraste, *Martyrium Sancti et Magni Martyris Anastasii Persae*. Patrologiae, Series Graeca Posterior, Tomus CXIV, Column. 773 – 812. Accurante J. – P. Migne. Parisiis 1864.

The Old Testament according LXX. Edition "Zoi." Athens 1986.

Union, Bulletin of "Christian Lodgings," May 2019.

Yahya Armajani, *CHRISTIANITY viii – Christian Missions in Persia*. Encyclopaedia Iranica, 1991.

Chapter 2

Legal Aspects of Mental Health in India

Khazin Munir[1*] and Mohammad Hussain[2]
[1]University of Kashmir, Jammu and Kashmir, India
[2]Law Department, University of Kashmir,
Jammu and Kashmir, India

Abstract

Mental Health Care (MHC) Act 2017 is published in the Gazette by the Government of India. It is aimed to intensify the human rights of persons with mental illness, and it gives them power regarding their health and treatment. Changes such as advance directives and nominated representatives are new, and review boards and responsibilities of government are clearly emphasized. And the act of attempt to suicide was decriminalized which was a big step as it was mentioned as a separate section in the Indian Penal Code. The objective of this Chapter is to give an overview of the Mental Health Care Act 2017 and the upcoming issues, its implementation and the challenges it must face in the coming future.

Keywords: mental health; mental health care; The Mental Health Act, 1987; The Mental Healthcare Act, 2017

Introduction

Mental health is nothing new to the world. With the existence of man, mental health has shown its existence. Every phase of life that a man cover is not

* Corresponding Author's Email: khazinmunir6@gmail.com.

In: Handbook of Uncertainty in Eurasian Forecasting (HEF)
Editors: Saeid Eslamian, Mir Bintul Huda and Faezeh Eslamian
ISBN: 978-1-68507-966-6

covered by him alone, his mind records every bit of it and justifies it in its own way. Mental health is a very underrated topic as people generally take it very casually but unfortunately giving less importance to it has led to the death of many by suicide. This issue leads not just to suicide only but divorce, family issues, issues at workplace etc. This one thing disturbs the whole life mechanism of a person.

Till the 17th century, all abnormal or mental behavior was supposed to be an act of the devil i.e., against God, the mentally ill were under the influence of evil spirits and described as witches. Gradually over the passing time, mental illness was considered abnormal behavior & mentally ill were considered socially intolerable and were incarcerated along with other criminals as they were also construed as criminals because their actions were against society, and, they could harm people around them. In the modern era, there was a shift from evil to ill. The mentally ill were called mad or insane positioned in special places called asylums. However, gradually these asylums became the place for human exploitation. Phillipe Pinel was the first psychiatrist to free the mentally ill from the asylum. Clifford Beers' work 'The mind that found itself' brought to light the treatment meted out to these people in asylums, resulting in a strong reaction to the plights of the mentally ill. This uproar resulted in the start of the 'mental hygiene' movement. In the 20th century, the work of Freud and 'B. F. Skinner & J. B. Watson gave a scientific combination of biological & social theories to explain the etiology of mental illness [1].

According to the *World Health Organization (WHO)*:" Mental health is a state of well-being in which an individual realizes his or her own abilities, can cope with the normal stresses of life, can work productively, and is able to contribute to his or her community." The WHO stress that mental health is "more than just the absence of mental disorders or disabilities." Peak mental health is about not only avoiding active conditions but also looking after ongoing wellness and happiness [2].

According to *Oxford Dictionary*, Mental health means, "A person's condition with regard to their psychological and emotional well-being [3]."

Black's law dictionary doesn't define mental health but defines Mental illness as "A disorder in thought or mood so substantial that it impairs judgment, behavior, perceptions of reality, it the ability to cop up with the ordinary demands of life [4]."

There are many reasons which can lead to mental disturbance or mental illness in a person, and some are as follows:

2019 onwards Coronavirus has been a big reason for depression and suicide amongst kids as well as adults, childhood abuse, trauma, or neglect, social isolation or loneliness, experiencing discrimination and stigma, social disadvantage, poverty or debt, bereavement (losing someone close to you), severe or long-term stress, having a long-term physical health condition, unemployment or losing your job, homelessness or poor housing, drug and alcohol misuse, domestic violence, bullying or other abuse as an adult, significant trauma as an adult, such as military combat, being involved in a serious incident in which you feared for your life, or being the victim of a violent crime, or living in a conflict zone (like Kashmir/Palestine/Burma/ Syria/Egypt, etc.) physical causes – for example, a head injury or a neurological condition such as epilepsy can have an impact on your behavior and mood. Although lifestyle factors including work, diet, drugs, and lack of sleep can all affect your mental health, if you experience a mental health problem there are usually other factors as well.

United Nations on Mental Health

Millions of people worldwide have mental health situations and an estimated one in four people globally will experience a mental health condition in their lifetime. Almost one million people die due to suicide every year, and it is the third leading cause of death among young people. Depression is the leading cause of years lost due to disability worldwide. Mental health problems, including alcohol abuse, are among the ten leading causes of disability in both developed and developing countries. Depression is ranked third in the global burden of disease and is projected to rank first in 2030 [5].

United Nations Convention on the Rights of Persons with Disabilities (CRPD)

The Convention on the Rights of Persons with Disabilities was passed by the General Assembly of the United Nations in December 2006. By February 2013, it has been signed by 155 countries and ratified by 127. It stated that citizens with a disability should enjoy a fair and peaceful society. It is one of the nine core human rights treaties of the UN. The overall purpose, stated in Article 1, is to "promote, protect and ensure the full and equal enjoyment of all human rights and fundamental freedoms by all persons with disabilities, and to promote respect for their inherent dignity." The elimination of discrimination by ensuring that rights may be enjoyed on an equal basis with others is a fundamental aim [6].

The word 'Disability' is not formally defined in the CRPD, allowing individual State Parties to define this term in accordance with their domestic law and in accordance with the prevalent conditions of their state.

According to United Nations the people with disabilities are characterized as, "Persons with disabilities include those who have long-term physical, mental, intellectual, or sensory impairments which in interaction with various barriers may hinder their full and effective participation in society on an equal basis with others." The use of the word 'include' in the statement above allows for a non-exhaustive description of 'disability' that is not settled; neither are the meanings of terms such as 'long-term' and 'impairments. It is accepted by the Committee on the Rights of Persons with Disabilities that people with a 'mental illness (referred to as having a 'psychosocial disability) fall under the Convention [7].

It states that mental illness is determined "in accordance with national and international accepted medical standards (including the latest edition of the International Classification of Disease of the World Health Organization) as may be notified by the Central Government." Additionally, the Act asserts that no person or authority shall classify an individual as a person with mental illness unless indirectly in relation to the treatment of the illness.

Indian Position on Mental Healthcare

Indian society has always been inclined towards the spiritual aspect of man and the conditions he is surrounded by. Man, and his relationship with God is of utmost importance in Indian society and maximum aspects depend on this fact only.

Maulana Fazulur-Lah Hakim, an Indian physician oversaw the first Indian mental asylum, i. e. Mandu Hospital opened by Mahmood Khilji (1436-1469) at Dhar, M. P. First lunatic Asylum, Bombay Asylum, was built in modern India in approximately 1750 A. D. at the cost of 125/-, no traces of it is present today. In 1794, a private lunatic asylum was opened at Kilpauk, Madras. The central mental hospital, Yerwada, Pune was opened in 1889. First asylum for insane soldiers was started at Monghyr, Bihar and was known as Monghyr Asylum (1795) [8].

One of the earliest Indian psychiatrists to explain the importance of health was Govindaswamy in 1948. He gave three objectives of mental health - regaining of the health of mentally ill person; prevention of mental illness in a vulnerable individual; and protection & development at all levels, of human society, of secure, affectionate & satisfying human relationships & in the reduction of hostile tensions in persons & groups (Govindaswamy, 1970). According to one aspect, put forward by Govindaswamy (1970) selfishness on the psychological side & starvation on the physical side are responsible for the disorganization of individuals & society. The second aspect stressed the importance of culture to understand personality functioning. Carstairs &Kapur (1976) & Chakraborty (1990) found the relation between social stress, modernization & occurrence of mental disorders. The third aspect is the use of traditional concepts of therapy e.g. Yoga by Patanjali' & fourth aspect is the importance of family in therapy [9].

During the reigns of King Ashoka, many hospitals were established for the mentally ill. A temple of Lord Venkateswara at Tirumukkudal, Chingleput District, Tamil Nadu, contains inscriptions on the walls belonging to Chola period. The inscription mentioned a hospital and a school. The hospital was named Sri Veera Cholaeswara hospital.

In India, traditionally persons with mental illnesses have been taken care of in the community by family members. During the rule by the British, many asylums were created and the maximum no. of mentally ill patients in these hospitals were britishers and British soldiers. Most of these hospitals continued after India got independence in 1947, and many new hospitals were constructed.

In the last three decades, many reforms have been initiated in the mental hospitals in India by involving the family members of the persons admitted and those attending the outpatient services. The strength of joint family, marriage, the close-knit community, greater tolerance of deviant behavior in the larger community, religion, and faith-based coping and healing have all contributed to a large number of persons with various mental disorders being taken care of in the community in India [10]. Internationally, the gradual closure of the mental hospitals has taken place due to the issues of frequent cases of maltreatment of patients, geographical and professional isolation of the institutions and the staff, failure of management and leadership, badly targeted financial resources, inadequate staff training, and not enough inspection and quality guarantee procedures. On the other hand, the mental hospitals in India have upgraded their hospitals by developing good training facilities, expanding outpatient and community services, and downsizing the inpatient units.

Involving families in taking care of the patients under the care of mental health services has been a unique contribution from India. It was initiated by Dr. Vidya Sagar in the 1950s at Amritsar Mental Hospital followed by the Mental Health Centre at Christian Medical College, Vellore, and All India Institute of Mental Health, Bengaluru, in the 1960s. Family members would be admitted along with the persons with mental illness to be a part of the care for the patient. This practice has been continued in most of the general hospital psychiatric units (GHPUs), which developed from 1960s onward in India. During the 1970s and 1980s, efforts were also made to understand the functioning of families with an ill person in the family and their needs [11].

Meanwhile, India was getting introduced to Mental health issues and a committee was established named as Bhore committee (in 1946) and on its recommendations All India Institute Mental Health was set up in 1954, which became the National Institute of Mental Health And Neurosciences in 1974 at Bangalore, which was launched in 1982 and later the District Mental Health Programme (DMHP) in a stepwise fashion from 1996 onward.

The national mental health program was launched with the aim to guarantee the availability and accessibility of minimum mental healthcare for all the people so everyone is prepared for the upcoming time, particularly the most vulnerable, underprivileged, and unnoticed sections of the society to promote the knowledge of mental health in general healthcare and in social development; and to promote community participation in the mental health service development. The focus of the NMHP was the amalgamation of mental health care with general, primary health care.

District Mental Health Programme was launched in 1996 with four districts, based on earlier experiences in Bellary District, Karnataka State. DMHP envisaged a community-based approach to the problem, which included training of mental health team at identified nodal institutions, increasing awareness about the mental health problems, and reducing the associated stigma provision of services for early detection and treatment of mental disorders in the community, and collecting information and getting experience at the level of community for future planning. It followed a more realistic and practical approach compared to the ambitious aims of the NMHP. Presently, the DMHP has been implemented in 241 districts of the country, and it is proposed to expand it to more districts. Currently, the emphasis is on a judicious balance between various components of the mental health care delivery system with clearly specified budgetary allocations. A plan for integration of NMHP with National Rural Health Mission (later renamed as the National Health Mission) was also developed [12].

The National Mental Health Survey of India in 2016 found that *"1 in 20 people in India suffer from depression,"* "productive age groups are affected most," "economic burden of mental disorders is huge" and that the treatment gap varies between 70% and 92% [13].

Different countries have a different approach to law toward their citizens. Sometimes a law is not formed as a separate legislature but is mentioned in general laws as a part of it.

In India, *the Mental Health Care Act, 2017* was passed on 7 April 2017 and came into force on 7 July 2018. On March 27, 2017, Lok Sabha in a unanimous decision passed the Mental Healthcare Act 2017 which was passed in the Rajya Sabha in August 2016 and got its approval from the Honorable President of India in April 2017. The opening paragraph of the Act states "An Act to provide for mental healthcare and services for persons with mental illness and to protect, promote and fulfill the rights of such persons during delivery of mental healthcare and services and for matters connected therewith or incidental thereto." Before this Act, the law which was prevalent was Mental Health Act, 1987 which was passed on 22 May 1987.

In British India Lunacy legislation included the Lunatic Removal Act 1851, the Indian Lunatic Asylum Act 1858, the Military Lunatic Act 1877, and some other Acts related to lunacy along with their amendments. The first law to govern mental health in India was the Indian Lunacy Act 1912, which was carved out of the English Lunatics Act 1845. The Indian Psychiatric Society suggested a draft in 1950, however, it was only given assent by the President in May 1987 and implemented in 1993 as the Mental Health Act 1987. Criticism of the Mental Health Act 1987 led to the Mental Health Care Bill 2013, which was finally passed as the Mental Healthcare Act, 2017.

The actual journey of the Mental Health Act, 2017 began from the *"Indian Lunacy Act, 1912"* which was basically the first law that covered the field of mental health and issues in India. But the focus of this Act was to protect and safeguard the common man and society from the mentally ill patients as they were considered dangerous and hazardous to society. This Act totally was silent over the rights of mentally ill people and neglected the privileges which should have been given to such patients and was this approach was highly criticized by the Indian Psychiatric Society.

This further gave birth to the *"Mental Health Act, 1987"* which finally came into existence in 1993. This act was an uplift of the Indian Lunacy Act,

1912 in terms of placing emphasis on care and treatment instead of custodial sentences, protecting the human rights, emphasizing on guardianship and management of the property of mentally ill people. It also provided measures for hospital admission on special circumstances. However, it was still not up to the expectations. Though having many positive features, the Mental Health Act 1987 had been the target of criticism right since its inception. It is alleged to be concerned mainly with the legal procedure of licensing, regulating admissions, and guardianship matters of PMI. Human rights issues and mental health care delivery were not adequately addressed in this Act.

It had many lacunas in it regarding some important aspects. It curtailed the liberty of such individuals which was a major disregard for human rights, it did not provide any relief to patients on rehabilitation and treatment plans after their discharge from hospitals. It also failed to take into consideration the financial, social, or emotional burdens to the patients as well as their caregivers.

All this made a pathway for the Mental health Act, 2017 which is a comprehensive Act and has been drafted keeping in mind the previous Acts and laws and the defects that had. The definition of mental illness has been clearly defined in Section 2(s) of the Acts as "a substantial disorder of thinking, mood, and perception, orientation, or memory that grossly impairs judgment, behavior, capacity to recognize reality or ability to meet the ordinary demands of life, mental conditions associated with the abuse of alcohol and drugs. But it doesn't regard mental retardation, a condition of arrested or incomplete development of the mind of a person, especially characterized by sub normality of intelligence, as mental illness [14]."

In 2018, The Mental Healthcare Act finally came into force. It includes:

Decriminalizing Suicide Attempt

In "section 115":

(1) Notwithstanding anything contained in Section 309 of the Indian Penal Code any person who attempts to commit suicide shall be presumed, unless proved otherwise, to have severe stress and shall not be tried and punished under the said code.
(2) The Appropriate Government shall have a duty to provide care, treatment and rehabilitation to a person, having severe stress and who

attempted to commit suicide, to reduce the risk of recurrence of attempt to commit suicide.

This shows how the formation of the act has allowed the sensitive care that has to be taken to such victims of suicide who are mentally stressed and unaware about their well-being, this act has allowed now to take special care to such cases wherein the victim has attempted suicide due to stress or mental illness and has provided provisions through which they cater to the needs of mentally unhealthy or unfit personnel.

Secures Human Rights of People with Mental Illness

It ensures that every person has the right to access mental health services. Such services should be of good quality, convenient, affordable, and accessible. It also protects such persons from inhuman treatment and provides access to free legal services and their medical records.

Advance Directives

This empowers a mentally ill person to have the right to make an advance directive of how they would want to be treated for their requisite illness and who should be their nominated representative. This directive must be vetted by a medical practitioner.

Mentally ill persons often lack the ability to make sound decisions and do not always have someone to speak for them on their behalf. In such a situation, treating physicians is the best to take decision because patients or their nominated representatives have limited knowledge of mental health and mental illness. Hence, from a physician's perspective, this new directive will lengthen the time of admission of mentally ill persons.

The Act further recognizes the right to community living; right to live with dignity; protection from cruel, inhuman, or degrading treatment; treatment equal to persons with physical illness; right to relevant information concerning treatment, other rights, and recourses; right to confidentiality; right to access their basic medical records; right to personal contacts and communication; right to legal aid; and recourse against deficiencies in the provision of care, treatment, and services.

Restricted Usage of Electroconvulsive Therapy

Few procedures which seems barbarian and clearly against human rights are prohibited exclusively. These procedures make mental healthcare seem to be an entirely gruesome experience, but these patients need to be aware that these procedures are forbidden and that they need not be scared and come forth with the treatment in a positive attitude.

I. Electro-convulsive therapy without the use of muscle relaxants and anesthesia,
II. Electro-convulsive therapy for minors,
III. Sterilization of men or women, when such sterilization is intended as a treatment for mental illness,
IV. Chained in any manner or form whatsoever.

Section 96 of the Chapter XIIV states: No psychosurgery shall be performed until:

I. The informed consent of the patient on whom surgery is being performed,
II. Approval from the concerned board to perform the surgery.

Free Quality Treatment for Poor and Homeless Persons

It also ensures that people with mental illness who are homeless or fall below the poverty line (BPL) (irrespective of whether they have a BPL card or not), will be given free quality treatment.

Ensures Right to Confidentiality

It also provides the right to confidentiality to people with mental illness in terms of their mental illness, healthcare, treatment, and physical healthcare.

It also ensures that photographs or any other information related to the patient will not be released in the media without their consent.

Special Provisions for Women

Certain special provisions have been made for women regarding their healthcare which includes not separating women from their children unless it is necessary.

Strict Punishments for Flouting

The punishment for flouting of provisions under this Act will be imprisonment up to 6 months or Rs. 10,000 or both. Offenders repeatedly violating provisions under this Act can face up to 2 years in jail or a fine of Rs. 50,000–5 lakhs or both.

Rehabilitation and Treatment

It also ensures that people with mental illness receive proper healthcare, treatment, and rehabilitation facilities in a way that does not impact or affect their dignity or rights [15].

Insurance to Mental Healthcare

Section 21(4)A of the Mental Healthcare Act, 2017, provides that there should be no discrimination between mental illnesses and physical illness and that every insurer shall make provision for medical insurance for treatment of mental illness on the same basis as is available for the treatment of physical illness. However, the majority of insurance companies exclude a large number of mental conditions from full coverage of the policy.

While the new act makes several provisions, it provides no guidelines or rules of implementation.

Focus

- Increase in public and professional awareness and intervention to reduce stigma and ease the burden of discrimination;
- Extend and strengthen existing systems of primary care to deliver health services for these disorders;
- Make cost-effective interventions available to those who will benefit;

- Mental health services should be accessible, equitable, and affordable;
- The government should downsize large psychiatric hospitals;
- Human resources for mental health must be systematically enhanced through both short-term and long-term strategies;
- There should be a national database of services and human resources available for mental health care in the country and this should be periodically updated;
- The State and Central Governments should follow a stepped care approach to mental health services;
- Aftercare rehabilitation and reintegration within the society;
- Mental health must be converged with the social, education, labor, and legal sectors. Translational research must be encouraged in all areas;
- Law review and reform need to occur periodically;
- Limitations imposed on the mentally ill in the area of insurance should be rectified;
- The mental health care of vulnerable groups like children, the elderly, and women subject to domestic violence should receive priority attention [16].

Conclusion

The Act is a reluctant approach toward a potential progressive law in mental health; adorned with all the precise provisions and rights guaranteed to the PMI, it falls short of combating the real issues head-on. Given the rampant lack of awareness about mental health, all the novel provisions for the PMI become tautological. Though the Act talks about sensitizing people about mental health, it doesn't provide promising methods for the same.

References

[1] SR Parkar, VS Dawani, JS Apte. 2001. History of psychiatry in India. *Journal of Postgraduate Medicine*, 1(47): 73-76. https://www.jpgmonline.com/article.asp?issn=0022-3859;year=2001;volume=47; issue=1;spage=73;epage=6;aulast=parkar.

[2] World Health Organization, https://www.who.int/india/health-topics/mental-health.

[3] Oxford Learner's Dictionaries, https://www.oxfordlearnersdictionaries.com/definition/english/mental-health.

[4] *Blacks Law Dictionary*. 2019. Thomson Reuters; 11th edition.

[5] United Nations, Department of Economic and Social Affairs, *Mental Health and Development*. https://www.un.org/development/desa/disabilities/issues/mental-health-and-development.html.

[6] Szmukler G, Daw R, Callard F. 2014. Mental health law and the UN Convention on the rights of persons with disabilities. *Int J Law Psychiatry*, 37(3): 245–252 https://www.ncbi.nlm.nih.gov/pmc/articles/PMC4024199/#:~:text=People%20with%20a%20mental%20illness,characterisation%20of%20persons%20with%20disabilities.

[7] United Nations, Department of Economic and Social Affairs, *Mental Health and Development*. https://www.un.org/development/desa/disabilities/issues/mental-health-and-development.html.

[8] SR Parkar, VS Dawani, JS Apte. 2001. History of psychiatry in India. *Journal of Postgraduate Medicine*, 1(47): 73-76. https://www.jpgmonline.com/article.asp?issn=0022-3859;year=2001;volume=47; issue=1;spage=73;epage=6;aulast=parkar.

[9] Isaac, M. Introduction. In: Chavan B S, Gupta N, Arun P, Sidana A, Jadhav S, editors. *Community Mental Health in India*. New Delhi: Jaypee Brothers Medical Publishers (P) Ltd.; 2012. p. xxxv-xli.

[10] Kumar, C.R., Nanda, P.B., Nitin, G. Recent developments in community mental health: Relevance and relationship with the mental health care bill. *Indian Journal of Psychiatry*, https://www.indjsp.org/article.asp?issn=0971-9962;year=2015;volume=31;issue=2;spage=153;epage=160;aulast=Chadda#ref2.

[11] Agarwal, S P, Ichhpujani R L, Shrivastava S, Goel D S. Restructuring the national mental health programme. In: Agarwal S P, Goel D S, Salhan R N, Ichhpujani R L, Shrivastava S, editors. *Mental Health: An Indian Perspective (1946–2003)*. New Delhi: Directorate General of Health Services, Ministry of Health and Family Welfare, Government of India/Elsevier; 2004. p. 119-22.

[12] Wikipedia. *Mental Health in India*. https://en.wikipedia.org/wiki/Mental_health_in_India#:~:text=Prevalence,-Further%20information%3A%20Suicide&text=The%20National%20Mental%20Health%20Survey,between%2070%25%20and%2092%25.

[13] Joshi, Sonal. India: Highlights of the brand new Mental Healthcare Act, 2017. *H.K. Acharya & Co.* https://www.mondaq.com/india/healthcare/805694/highlights-of-the-brand-new-mental-healthcare-act-2017.

[14] H.K. Acharya & Company. *Highlights of the Brand New Mental Healthcare Act, 2017.* http://www.hkindia.com/news_letter/article/3/litigation_article-3.html?utm_source=Mondaq&utm_medium=syndication&utm_campaign=LinkedIn-integration.

[15] YLCube, *Mental Healthcare Act, 2017- An Analysis*. https://nitimanthan.in/blog-posts/blog-niti-manthan/2021/03/21/mental-healthcare-act-2017-analysis/.

Part II: Climate Change and Ecological Changes (CCEC)

Chapter 3

Natural Disasters

Never Mujere*
Department of Geography Geospatial Sciences and Earth Observation, University of Zimbabwe, Harare, Zimbabwe

Abstract

Natural disasters threaten ecosystems, human life, and economic development greatly. However, there is limited research focusing on natural disasters in arid and semi-arid regions of southern Africa. This chapter explains the causes and effects of natural disasters and the disaster management cycle in general. It then zeros into specific major natural disasters occurring in Zimbabwe. It also highlights the framework for understanding natural disasters. The discussion dovetails on the management of natural disasters in the context of civil protection in the country. The Civil Protection Unit is a national disaster response institution that is actively rolling out education and awareness-raising programs so that vulnerable people are more prepared to face disasters and to react adequately when they occur. The Unit, though under-resourced, is actively designing emergency and evacuation plans such as escape routes to safer places in times of flooding.

Introduction

Natural disasters such as floods, droughts, tsunamis, tropical cyclones, and earthquakes are characteristically extreme, rare, less frequent, high magnitude, difficult to predict and cause enormous damage to property, environment and

* Corresponding Author's Email: nemuj@yahoo.co.uk.

In: Handbook of Uncertainty in Eurasian Forecasting (HEF)
Editors: Saeid Eslamian, Mir Bintul Huda and Faezeh Eslamian
ISBN: 978-1-68507-966-6

life. They are continuing to reign havoc globally, while little attention has been given to addressing them. Urgent multidisciplinary action is required to deal with large-scale risks and pressing challenges associated with natural disasters (Wisner, 2020). Preventing disasters contributes to the achievement of the UN Sustainable Development Goals (SDGs) especially SDG Number 6 and Agenda 2030. In many countries, there is a need to create qualified technical capacity among stakeholder groups to develop and implement sound policies and strengthen disaster governance regimes at national and regional levels. For this, disaster modeling can be a powerful instrument that might also open new development perspectives and present attractive opportunities for coping with looming environmental threats.

The term disaster owes its origin to the French word '*desastre*' which is a combination of two words *'des'* meaning bad and *'aster'* meaning star. Thus, the term refers to a 'bad or evil star' (WHO, 2002). The United Nations defined Disasters as 'A serious disruption of the functioning of a community or a society causing widespread human, material, economic, and environmental losses which exceed the ability of the affected community/society to cope using its own resources (WHO, 2002). A disaster results from the combination of hazard (i.e., triggering event with the potential to cause damage), vulnerability (i.e., susceptibility of a community to be damaged by a particular hazard) and insufficient capacity or measures to reduce the potential chances of risk. A disaster occurs when a hazard impacts a vulnerable population and causes damage, casualties and disruption. Thus, a disaster takes place only when hazards occur in an area with vulnerable elements. But with greater capacity of the individual, community and environment to face these disasters, the hazard impact reduces.

Collaboration and cooperation are crucial to disaster risk reduction. Local, regional, and international organizations and institutions all have a role to play in disaster management. Civil society, including volunteers and community-based organizations, the scientific community, the media, and the private sector, are all vital stakeholders. Wisner (2020) states that the need to manage disasters has led to the ratification of international agreements such as the Hyogo Framework for Action (2005-2015) and the Sendai Framework for Disaster Risk Reduction (2015-2030). The primary objective Hyogo Framework for Action was to prepare and protect against natural disasters thereby creating a safer environment. Different stakeholders were engaged in reducing the risks and losses associated with disasters. Stakeholders are to be drawn primarily from the public and private sectors and should include communities and international agencies. The priorities of action of the Hyogo framework were to ensure that disaster risk reduction is a national and a local priority with a strong institutional basis for implementation. It calls for strong national and local commitment is required to save lives and livelihoods threatened by natural hazards. Also, it spells out that disasters must be considered in public and private sector decision-making in the same way that environmental and social impact assessments are currently required. They must also allocate sufficient resources to support and maintain them. The following is an extract indicating the importance of the Hyogo Framework for Action.

> "Climate change is expected to cause more severe and more frequent natural hazards. As our cities and coasts grow more vulnerable, these hazards can lead to disasters that are far worse than those we have seen to date. We have a moral, social, and economic obligation to build resilience by 2015. Implementing the Hyogo Framework for Action will also help us reach the Millennium Development Goals."
>
> *Ban Ki-moon, Secretary-General of the United Nations.*

The Sendai Framework for Disaster Risk Reduction was adopted at the third United Nations conference on disaster reduction held from 14 to 18 March 2015 in Sendai, Japan. The goal of the Framework is to prevent new and reduce existing disaster risks through the implementation of integrated and inclusive social, cultural, economic, political, legal, health, educational, environmental, technological, and institutional measures that prevent and reduce hazard exposure and vulnerability to disaster, and increase preparedness for response and recovery, and thus strengthen resilience (Wisner, 2020). As a result of these and other innovations, national governments are increasingly aware of the dangers and costs imposed by inappropriate land use and environmental practices.

This work is structured as follows. Section 1 introduces the chapter. Section 2 details the various types of natural disasters. Section 3 presents the disaster cycle. Section 4 describes the common natural disaster occurring in Zimbabwe. Section 4 discusses the role of the civil protection unit in Zimbabwe. Section 5 summarises the useful and important points in this paper.

Types of Natural Disasters

Disasters vary in space and over time in terms of their origin, magnitude, speed of onset, duration, frequency, areal extent and areal reliability. High magnitude disasters pose sufficient threats to the human systems whereas low magnitude events pose low impacts on life and property. For example, the magnitude of an earthquake can be measured using a Richter Scale,

The speed of disaster onset refers to the length of time between the first appearance or warning of an event and its peak. Drought and soil erosions is slow-onset or creeping disasters while floods and earthquakes are rapid-onset disasters.

Duration of hazards refer to the period over which it occurs. Droughts occur over seasons or years, floods occur in days or weeks whereas tornados take place in minutes or hours.

Frequency of disaster occurrence refers to how often an event of a given magnitude may be expected to re-occur. This is statistically characterized using the return period' or a recurrence interval. An event with a return period of 10 years has a 10% chance to occur in a year. For example, droughts in southern Africa have been shown to occur after every 10 years.

Table 1. Types of natural disasters

Type of Natural Disaster	Explanation	Examples
Hydrological	Caused by the occurrence, movement, and distribution of the surface and subsurface freshwater and saltwater.	Streamflow floods, flash floods, hydrological droughts, glacial lake outbursts, flood, landslide and wave action.
Meteorological or climatological	Caused by short-to long-lived, micro- to macro-scale extreme weather and atmospheric conditions ranging or lasting from minutes to days, intra-seasonal to multi-decadal climate variability.	Heat waves, convective storm, fog, extreme temperature, cold waves, hailstorms, wind storms, thunderstorms, lightning, airburst. tornadoes, tropical cyclones, meteorological droughts, and wildfires.
Geophysical or geological or geomorphological	Originate from solid earth.	Earthquakes, rock falls, mass movement, volcanic Activity and wildfires.
Biological	Vector-borne diseases and plagues caused by the exposure to living organisms.	Animal incident, malaria, insect infestation, armyworm, anthrax. ebola, new castle.
Extra-terrestrial	Caused by asteroids, meteoroids, and comets as they pass near earth, enter the earth's atmosphere, and/or strike the earth, or change in interplanetary conditions that affect the earth's magnetosphere, ionosphere, and thermosphere.	Meteorite and other extra-terrestrial impacts.

Source: Adapted from Shi (2019).

The areal extent of a disaster refers to the area affected by the natural disaster. An avalanche occurs within a short and narrow belt of the landscape, whereas a drought or flood affects a large geographical space extending several kilometers.

Areal reliability is the predictability of natural hazards as they occur in each area. Predictable or point sources include volcanic eruptions occurring in fixed points. Floods occur in channels and flood plains. Tropical cyclones occur over several paths. Earthquakes are rarely predictable as their locations are less fixed and hence are non-point sources.

With regards to origins or drivers, natural disasters can be grouped into five types namely, meteorological, hydrological, geological and geomorphological, biological, and disease hazards (Shi, 2019). Table 1 highlights the five broad categories of natural disasters.

Disaster Management Cycle

The disaster management cycle illustrates the ongoing process by which governments, businesses, and civil society plan for and reduce the impact of disasters, react during and immediately following a disaster, and take steps to recover after a disaster has occurred. The four stages in disaster management are mitigation, preparation, response, and recovery (Figure 1). The whole disaster management cycle depends on the incorporation of appropriate measures in national and regional development plans. In addition, its effectiveness depends on the availability of information on disasters, emerging risks and the countermeasures to be taken, and on the degree to which government agencies, non-governmental organizations and the public can make use of this information.

Mitigation activities eliminate and prevent the probability of disaster occurrence. They also reduce the effects of unavoidable disasters. Mitigation measures include building codes; vulnerability analysis updates; zoning and land use management; building use regulations and safety codes; preventive health care and public education. Effectiveness of mitigation depends on the availability of information on disasters, emerging risks, and the countermeasures to be taken.

The goal of emergency preparation programs is to achieve a satisfactory level of readiness to respond to any emergency through programs that strengthen the technical and managerial capacity of governments, organizations, and communities. These measures are described as logistical readiness to deal with disasters and are enhanced by having response mechanisms and procedures, rehearsals, developing long-term and short-term strategies, public education and building early warning systems. Preparedness take the form of ensuring that strategic reserves of food, equipment, water, medicines and other essentials are maintained in cases of national or local catastrophes. During the preparedness phase, governments, organizations, and individuals develop plans to save lives, minimize disaster damage, and enhance disaster response operations. Preparedness measures include preparedness plans; emergency exercises and training; warning systems; emergency communications systems; evacuations plans and training; resource inventories; emergency personnel contact lists; mutual aid agreements; and public information and education.

The aim of disaster response is to provide immediate assistance to maintain life, improve health and support the morale of the affected population (WHO, 2002). Such assistance may range from providing specific but limited aid, such as assisting refugees with transport, temporary shelter, and food, to establishing a semi-permanent settlement in camps and other locations. It also may involve initial repairs to damaged infrastructure. The focus in the response phase is on meeting the basic needs of the people until more permanent and sustainable solutions can be found. Humanitarian organizations are often strongly present in this phase of the disaster management cycle.

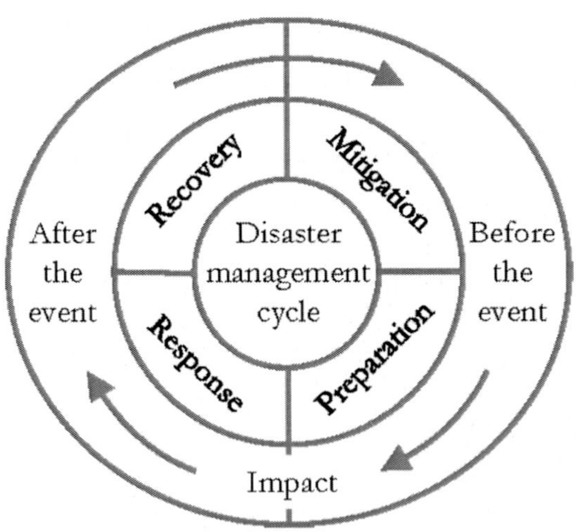

Figure 1. The disaster management cycle.

Recovery activities occur as the emergency is brought under control, and the affected population can undertake a growing number of activities aimed at restoring their lives and the infrastructure that supports them. There is no distinct point at which immediate relief changes into recovery and then into long-term sustainable development. There are many opportunities during the recovery period to enhance prevention and increase preparedness, thus reducing vulnerability. Recovery activities can be short-term or long-term and they continue until all systems return to normal or better. Recovery measures include returning vital life-support systems to minimum operating standards; temporary housing; public information; health and safety education; reconstruction; counseling programs; and economic impact studies. Information resources and services include data collection related to rebuilding, and documentation of lessons learned.

Response mechanisms are employed to save lives and minimize immediate impacts. During recovery, activities and services are restored. Prevention aims to minimize future disasters and preparation strategies enhance the preparedness of communities. The principle of 'building back better,' which is the use of recovery, rehabilitation, and reconstruction phases after a disaster aims to increase the resilience of nations and communities. It aligns with the characteristics of ecological resilience.

Controlling disasters aims at reducing their devastating effects. In some cases, disasters can be turned into a benefit, that is, living with disasters. This occurs when people take advantage of disasters to generate incomes and improve livelihoods. For example, during flooding, some farmers may grow rice and bananas. Effective mitigatory or coping measures call for a sound understanding of the drivers of disasters through supporting monitoring and research programs. Statistical analyses use past records to predict the occurrence of disasters at various return periods and magnitudes are commonly employed. When dealing with disasters structural and non-structural measures are employed. Structural hardware measures include the building of critical infrastructure to deal with hazards and their associated disasters. Non-structural or software measures involve the absence of physical infrastructure in place to manage disasters. They comprise disaster risk mapping, early warning and forecasting systems, insurance services, emergency and evacuation plans catchment management, education, and awareness-raising.

Natural Disasters in Zimbabwe

Zimbabwe is a semi-arid arid developing country extending from 15° 17^1 to 22° 24^1 South and from 25° 14^1 to 33° 04^1 East in southern Africa. It is among other countries in the SADC region and the world at large that has traditionally been vulnerable to natural disasters on account of its unique geo-climatic conditions. Floods, droughts, cyclones, wildfires, landslides, heat waves, pests, and diseases have been recurrent phenomena. In recent years have seen several people being left homeless and infrastructural damages due to a dramatic increase in cyclones-induced floods. Such hydro-climatic disasters have had a direct impact on the poor, incapacitated and marginalized communities of the country. This section discusses issues of floods, cyclones, wildfires, and droughts in Zimbabwe.

Floods

Naturally flooding happens when an area is inundated by a sudden rise in water levels due to heavy rainfall. In Zimbabwe, both streamflow floods and flash floods are experienced. Positively, floods brought good grazing to affected areas but grazing riparian to flooded rives was washed away. Also, they boost the water bodies in regions experiencing droughts such as the west, and south eastern parts of the country.

Fluvial floods occur when river levels rise and burst or overflow their banks, inundating the surrounding land that forms the river's floodplain. This can occur in response to storms having higher than normal rainfall totals and/or intensities, seasonal strong weather systems such as cyclones.

Flash floods develop when heavy rainfall occurs suddenly, particularly in mountainous and urban river catchments, although they can occur anywhere. The rate at which river water levels rise is very rapid and the flood forms quickly. High amounts of localized rainfall, rapid flood formation, and high-water velocities can be particularly threatening populations at risk and highly destructive. Challenges in the management of flash floods include the short preparation time to activate flood alerts and emergency response, and the sudden nature of the phenomenon, which often catches the population at risk by surprise.

Floods in Muzarabani District

Muzarabani district extends from 30°45''E to 31°20''E and 16°00''S to 16°30''S to the north of Zimbabwe (Madamombe 2004). It is along the flood plains of the Zambezi River, with Lake Kariba upstream and Lake Cabora Bassa downstream at the confluence of the Msengezi and Zambezi rivers about 400 m.a.s.l. Droughts have forced people to settle on the flood plains and along river banks. Floods have become a perennial problem in the district (Ncube-Phiri et al., 2014). Besides, backflow from Cabora Bassa dam and releases from Kariba Dam, flooding in the area is a natural result of increased flow at river confluences and inflows into the Zambezi River.

The effects of flooding in the affected area include loss of life and property. This includes damage to agriculture, transport, shelter, injury and death of animals and people. The emergence of water-borne and water-related diseases and financial burden to the country in dealing with flooding. During the December 2007 floods, a total of 1,000 households were affected, with 400 of them losing almost all their belongings (IFRC, 2007). The flooding also results in outbreaks of water-borne diseases such as malaria and cholera, further subjecting people to huge losses.

The Zimbabwe National Water Authority has produced a flood zoning map that defines flood-prone areas and the risks associated with different zones along the river. This allows for better preparation of emergency and evacuation plans (Figure 2).

Flood plain and catchment management are activities underway in river catchments as most reservoirs and rivers are silting up. This is due to poor catchment management such as poor agricultural practices, deforestation, and increased human settlements. Thus, responsible institutions and stakeholders are involved in de-silting water courses and educating farmers on environmentally friendly farming practices.

The Department of Meteorological Services is actively participating in the national early warning system for the nation and for SADC countries in general. The national early warning system collects, processes, analyses and disseminates a lot of data which is used for various medium and long-term planning purposes, whose aim is to reduce the risk of disasters. The country has developed flood warning systems comprising rainfall or hydrometric and communication systems to minimize the damages caused by floods by (i) providing a forecast days or hours in advance of the areas that are going to be flooded, water levels and eventually water velocity and (ii) allow series of measures and interventions, for example, reservoir operation, strengthening of dykes, removal of equipment and evacuation of population to be done in time. The forecast is derived from mathematical models, using rainfall, water levels, and discharges in upstream sections as inputs. Tools used to forecast floods include the use of statistical models that use past records to predict future conditions and the use of global climate change models. However, there is a need for capacity building particularly in the country to enable the national institutions to collect real-time accurate and relevant data, analyze and understand the results and make correct predictions and disseminate the information.

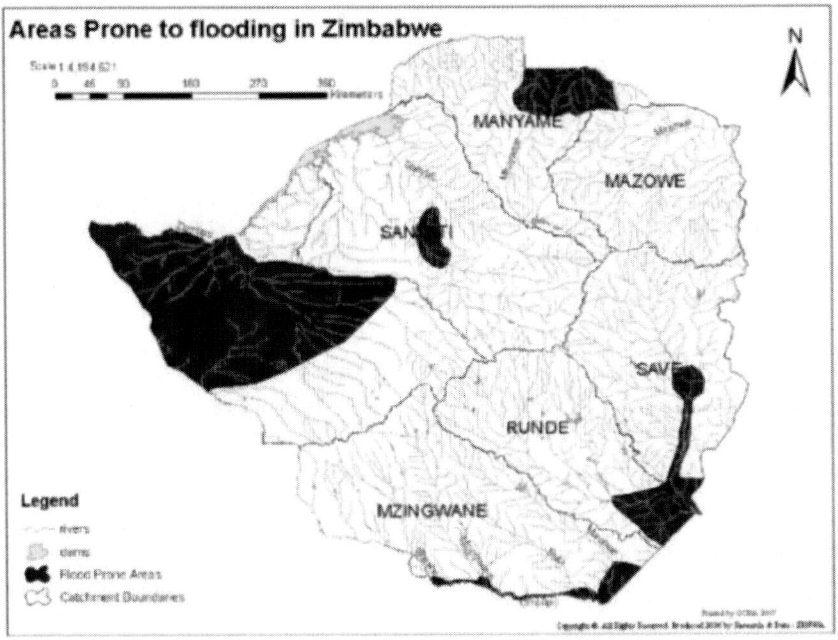

Source: www.zinwa.co.zw (Accessed 12 August 2020).

Figure 2. Areas prone to flooding in Zimbabwe.

Cyclones

Cyclones that hit the country in 2000 were Eline (23 February 2000), Huda, Japhet (March 2003), Flavio. Dineo (February 2017), Idai (15 March 2019), Kenneth (25 January 2019), Chalane (31 December 2020-3 January 2021) and Eloise (22-25 January 2021).

Cyclone Eline

Cyclone Eline hit the eastern and southern parts of Zimbabwe from 23 to 25 February 2000. The Government of Zimbabwe, through the president officially declared a state of disaster on 25 February 2000. In December 1999 MSD issued info predicting high probabilities of normal to above normal rainfall in southern Africa including Zimbabwe during the period January-March 2000. The Zimbabwe MSD announced cyclone movement from the east of Madagascar into Mozambique and then into Zimbabwe. The information was disseminated through newspapers, the radio, TV, pamphlets and CPU meetings. However, the concept of a 'cyclone' was very abstract to many since very few of the local authorities and farmers had experienced this disaster before. Also, the majority did not know what measures to put in place and could not take the warnings seriously.

The areas affected by the cyclone received well above normal rainfall during the period. The cyclone had a devastating effect on infrastructure such as roads, bridges, irrigation schemes, dams, electricity and telephone lines. The cyclone affected 217,900 people (2.1% of national) caused 91 deaths, 357 injuries, damaged 59,187 homes and affected approximately 2.7 million people.

Almost 2,239 cattle died, 13,149 sheep and goats died, and 1,180 donkeys and 18,320 fowls were washed away and died. Total livestock losses are estimated at US$753,500. Livestock infrastructure including kraals, dip tanks, cattle fences, and holding pens was damaged at a cost of US$29,000. The extended rainy season increased incidences of tick-borne diseases such as *Anaplasmosis* (gall sickness), *Babesiosis* (Redwater), *Cowdriosis* (Heartwater), Theileriosis (January disease), Ephemeral fever, Lumpy Skin disease, Black leg, and Anthrax. Common poultry diseases were Newcastle and Fowlpox in the affected areas.

In the affected areas, the incessant rains caused water-logging, crop lodging, and leaching of soil nutrients resulting in crop yields reduction by up to 60%. Continuous wet conditions hampered weeding, chemical spraying and harvesting. Grey leaf spot increased in late-planted maize.

A total of 28 dams were damaged of US$12,974,500 for rehabilitation costs. Rainfed farming lost 344,060 ha of cropped land. A total of 33 irrigation schemes with an area of 4,230 ha were damaged thus affecting 6,380 plotholders. Irrigation rehabilitation required US$2,042,047.

The timber industry lost almost 20% of its trees due to floods. Bridges and roads were destroyed. This means that internal and external markets (for timber and coffee) became inaccessible

Cyclone Idai

Cyclone Idai hit the country on 15 March 2019 and lasted for 2 days. The storm and subsequent flooding and landslides left traumatic experiences in nine districts of the country. It was deadly and similar to the one experienced in the 1940s. The estimated cost of the cyclone stood at a range of \$548–\$622m and \$557–\$767m for emergency response and recovery respectively (Chatiza, 2019).

The cyclone affected 270,000 people. It left 347 dead, 51,000 displaced, 354 hospitalized and 183 injured. About 344 people were reported missing and some believed to have been carried by flood waters into neighboring Mozambique (Chatiza, 2019).

Livelihoods were disrupted and 140 schools were affected. Housing, health, irrigation, and other agriculture facilities were damaged, as were forests and protected areas. Arable land was rendered unusable and at least 348 cattle, 17,000 chickens, and 222 goats and sheep were lost, alongside losses of stored cereals (Chatiza, 2019).

The cyclone damaged 250 boreholes, 13 health care centers, 18 urban and peri-urban water supply systems. Most parts of the eastern districts of Chiminimani and Chipinge were completely cut off from the rest of the country as roads, bridges, and communication systems were destroyed (Tsuro Trust, 2020). Almost 1,500 km of the road network was destroyed. Thus, affecting access to essential services such as markets, clinics and schools (Chatiza, 2019). Due to poor visibility and the risk of crashing in mountainous terrain which is interspersed by deep valley floor, helicopters could not fly into the affected areas. There was a five-day period in which the affected areas were virtually inaccessible and local action, though of significance, was undocumented. Figure 2 shows the destruction caused by Cyclone Idai.

In response to cyclone disasters, a wide range of activities was put in place. These include (Chatiza, 2019; Tsuro, 2019): establishing seven relocation centers covering 2,889 ha; counseling and psychosocial support, casework and child protection support, orphan care and family tracing; distributing food and non-food item distribution educational assistance to children, provision of water, health and sanitation facilities; vocational skills training and support and capacity building of those involved in disaster response.

Cyclone Idai exposed the nation's lack of resources and adequate horizontal-vertical connection among strategic institutions. Stakeholders fell short to gather and disseminate effective information to facilitate appropriate decision-making. Disaster management is still lacking. This explains why the institutional arrangements fall short to deal with disasters.

Figure 3. Some trails of destruction left by the cyclone in Chimaniani district. Source: Tsuro Trust (2020: 28).

Drought

A drought is an event that occurs when there is a shortage of water due to poor rainfall for an extended period. The period of concern is the rainy or farming season. It is a creeping phenomenon that slowly sneaks up. It has huge adverse effects and gives rise to significant losses in the economy, ecology, and environment. It causes crop losses, degradation and desertification, urban water supply shortages, and forest fires. In Zimbabwe, 60% of the households derive their livelihoods from agriculture (Ministry of Agriculture, 2000; Tsuro Trust, 2020). Thus, rainfall is a limiting factor in achieving food security.

The El Nino-Southern Oscillation (ENSO) often leads to drought conditions in the country. Past El-Nino years include 1979, 1982, 1992, 1999, 2006 and 2015. The 2015/16 season experienced a very strong El Nino. It was not a good season with a reduction in cereal production of 21%. The country. The country received below-normal rainfall accompanied by high temperatures leading to reduced crop yields, cattle deaths, and low dam levels due to fewer inflows. Sectors affected include food security, livelihoods, agriculture, livestock, nutrition, health and water, sanitation, and hygiene.

Drought and monitoring have become important activities to provide an early alert of drought occurrence. Remote sensing-based drought indices commonly used to monitor droughts include the Normalised Difference Vegetation Index (NDVI). NDVI values above 0.5 indicate healthy vegetation. While values below 0.5 show unhealthy vegetation or low vegetation cover.

Drought forecasting and prediction make use of models fed by historical and seasonal climatic and atmospheric data. During forecasting, real-time meteorological monitoring and weather forecasts are used to predict the probability of a drought occurrence in a region or area of interest. Drought mitigation, preparedness, and response comprise the appropriate measures and actions aimed at reducing the vulnerability to drought and reducing the impacts of droughts.

Wildfires

Wildfires are also known as bushfires or veld fires in Southern Africa. They occur in rural and urban areas. Natural fires in Zimbabwe that experience the savanna climate comprising vast grasslands with scattered trees are prevalent in the hot and dry summer season. Naturally, bush fires are caused by lightning, geothermal heat, excessive heat or high temperatures, volcanoes, and rock weathering. Wildfires spread faster uphill than downhill, with the wind rather than against the wind, where there is high biomass, small sources of fuel, and where the vegetation is intertwined.

Wildfires cause socio-economic and environmental effects in the country. On a positive note, they improve soil fertility, and germination process of lilies, help during the reproduction and life cycle of pine trees, decrease disease insects, rid invasive species in an ecosystem, and provide lupine food for blue butterflies and create new habitat zones.

These socio-economic impacts of wildfires include loss of agricultural produce; reduced food availability for both humans and animals; destruction of grazing land; loss of shelter and property causing trauma and stress; reduction of mental well-being and respiratory diseases such as asthma and bronchitis. The disastrous environmental effects of fire comprise biodiversity loss; habitat loss; increased air pollution and carbon dioxide emissions; land degradation; the acid rain and enhance floods.

During the 2018 fire season, the country recorded a total of 1,595 incidences resulting in 1,190,174.89 ha of land being destroyed while 2,705 incidences and 1,268,534.31 ha were burnt in the 2017 fire season. This translates to a 6.2% reduction in the area destroyed by veld fires and a 16.7% reduction in fire incidences. Most of the fires occur in the resettlement areas (62,2756.09 ha), communal areas (137,983.10 ha) and protected areas (gazetted forests, 55,376.25 ha, and national parks, 81,636.11 ha) continue to experience significant fires (EMA, 2019).

The 2018 fire season has seen the country losing 5 lives, property worth US$41,300, plantation valued at US$257 500, agricultural produce, farm equipment, and livestock valued at US$59,700. The 2018 fire season cost the nation a total of US$358,500 (EMA, 2019).

Strategies employed to mitigate the effects of wildfires include community education and awareness-raising, construction of fireguards, mapping fire prone areas, grass harvesting, and hay bailing. There are education and awareness initiatives to capacitate and forearm communities in the event of fire outbreaks. The government and civil society train communities in fire preventive structures such as controlled burning, using fire breaks, fire guards, and providing fire fighting pumps. Communities are also given fire beaters to fight fires and encouraged to report any fire incidences.

Thatch grass harvesting and hay bailing is a strategy used to prevent fire damage. During the 2018 fire season, a total of 368,000 hay bales were made countrywide with an estimated value of US$552,001.50. It is after realizing the value of the forests produce and the environment that communities tend to fully protect the environment from veld fires (EMA, 2019).

Mapping fire-prone areas and monitoring fires include the creation of fire vulnerability maps from remotely sensed fire data in a GIS. The maps are useful for determining fire hazard zones, monitoring fires, visualizing areas mostly affected by fires, and fire frequency. Mapping burnt areas are critical in damage assessment. The application of remote sensing GIS in fire monitoring helps in mapping out areas and identifying those most prone to burning. GIS supports planning, preparedness, mitigation, response, and incident management. Indices such as the Normalized Burnt Ratio (NBR) can be estimated to determine burnt and unburnt areas. An NBR value greater than 0.25 indicates burnt areas and vice versa. This allows a much faster and more effective response when and if a fire occurs. Mapping out these areas helps a community in knowing the hotspot areas, that is, the areas which are most prone to natural fires and this, in turn, will help in knowing where not to build their homes or place their crops and livestock which reduces any foreseen damage to their surroundings.

The Zimbabwe Civil Protection Unit

Civil protection describes all humanitarian measures taken to protect populations, their environment `and their natural inheritance against accidents and disasters of every kind. The government has legislated for the coordination of preparedness planning for emergencies and disasters through the Civil Protection Act Chapter 10.06 of 1989. The Minister of Local Government Rural and Urban Development is responsible for coordination as enshrined in the Civil Protection Act No. 5 of 1989.

Special powers include establishing, coordinating and directing activities of the private and public emergency services. It develops guidelines for action and maximum use of resources since disaster mitigation requires a multisectoral and interdisciplinary approach. It is also responsible for commandeering resources for the purpose of mitigating disasters and declaration of a state of disaster by the country's president. The National Civil Protection Fund is used to develop and promote civil protection activities throughout the country.

The CPU hierarchy transcends from national to district levels. At the national level, the minister, secretariat, and committee with multisectoral representatives are key actors. The provincial-level comprises a provincial administrator and a committee with multisectoral representatives. At the district level, the unit consists of a district administrator and a committee with multisectoral representatives.

Weaknesses of the CPU

- Its structure rarely caters to the inclusion of the community in its operations and yet communities first fight the disaster before external assistance using their indigenous knowledge;
- Politics often overrides technocrats. This is where disaster statistics are politically manipulated to make the situation seem to be under control while in reality, it would be not;

- Zimbabwe is inadequately resourced to deal with natural disasters. Its capacity is dependent on NGOs and the private sector which often delay chipping in for political, institutional (i.e., due to lack of coordination between the CPU) and NGOs, and corruption reasons;
- Lack of coordination between the CPU and NGOs often results in duplication of efforts and delay in disaster response;
- The UNIT has a critical human recourses shortage and those available occupy various offices and lack skills in dealing with disasters. The knowledge gap of natural hazards at the community level implies that the CPU is ill-resourced in doing its work to educate and raise awareness among people;
- It mostly uses the radio and television to warn people of impending disasters. Unfortunately, disaster-prone areas are inaccessible due to poor road and telecommunication networks;
- The process of declaring a hazard as a disaster is too bureaucratic, thus delaying response actions by NGO and other institutions;
- The existing CPU legislation is reactive rather than proactive.

In order to enhance sustainable disaster management, legislation on disaster risk management should place emphasis on prevention and mitigation; as well as recovery, rehabilitation and reconstruction, rather than mere response. The government should adequately fund the CPU to be disaster prepared in terms of having an emergency reserve fund, all-weather equipment and aircraft considering that the world is experiencing climate changes, health facilities, warning systems and logistical equipment. Capacitation should extend to training in proper planning, and stores management among other disciplines. There is a need for a robust early warning system that is accessible and easily understood at grassroots levels. The focus should be on all other disasters not only concentrating on those which are climate-related. Community involvement and participation in disaster risk management need to be enhanced.

The application and use of forensic science and remote sensing in disaster management can give information in real-time and data for accessible areas at a low cost.

Conclusion

Natural disasters are catastrophes caused by natural phenomena such as typhoon, flood, heavy rainfall, extreme wind, wind wave, tsunami, heavy snowfall, thunder, drought, earthquake, yellow sand, massive tide, volcanic activity, collision, and clash with natural objects in the universe including asteroids and meteors, among others. The chapter discusses disaster management aims to avoid, lessen or transfer the adverse effects of disasters through activities and measures for prevention, preparation, response, and recovery.

References

Environmental Management Agency, EMA. (2019). *Zimbabwe 2018 Fire Season Report.*

International Federation of Red Cross and Red Crescent Societies. 2007. *Zimbabwe: Muzarabani Floods.* http://www.ifrc.org/docs/appeals/07/MDRZM0 02a.pdf.

Madamombe, E. K. (2004). Zimbabwe: Flood Management Practices – Selected Flood Prone Areas Zambezi Basin. *WMO/GWP Associated Programme on Flood Management.* Zimbabwe National Water Authority, Harare.

Manatsa, Desmond & Chatiza, Kudzai & Mushore, Terence Darlington & Mudavanhu, Chipo & Sakala, et al. TSURO Trust. (2020). *Building Resilience to Natural Disasters in Populated African Mountain Ecosystems: The Case of tropical Cyclone Idai in Chimanimani*. Graphtec.

Ministry of Agriculture Zimbabwe. (2000). *Assessment of the Impact of Cyclone Eline (February 2000) on Food, Agriculture and Natural Resources Sector in Zimbabwe*. https://fews.net/sites/default/files/documents/reports/1000050.pdf.

Ncube-Phiri, S., Mudavanhu, C. and Mucherera, B. (2014). The Complexity of Maladaptation Strategies to Disasters: The case of Muzarabani, Zimbabwe. Jàmbá: *Journal of Disaster Risk Studies*, 6(1), 11pp. http://dx.doi.org/10.4102/jamba.v6i1. 145.

Shi, P. (2019). *Disaster Risk Science*. Beijing Normal University Press, Beijing. China. Series, https://doi.org/10.1007/978-981-13-6689-5_1.

Wisner, B. (2020). Five Years Beyond Sendai—Can We Get Beyond Frameworks?. *Int. J. Disaster Risk Sci.* 11, 239–249. https://doi.org/10.1007/s13753-020-00263-0.

Wisner, B., Adams, J. *Environmental Health in Emergencies and Disasters: A Practical Guide*. WHO. (2002).

Chapter 4

Drought Monitoring Indices

Nagilla Nagadatt Sharma[1,‡], Mahesh Ramesh Tapas[2,*], and A. Uday Kumar[3,†]

[1]National Institute of Technology, Warangal, India
[2]East Carolina University, Greenville, NC, USA
[3]Central Groundwater Board, Faridabad, Ministry of Jal Shakti, Government of India

Abstract

Several anthropogenic activities and climate changes are affecting watershed characteristics. The increased prolonged period between rainfall events is severely affecting the watershed. Groundwater depletion, non-conjunctive use of surface water and groundwater, and increasing water demand with a rising population are stressing a catchment's water demand and supply. Such a deficit in the water supply stresses plant productivity and drought situations in the watershed. For policymakers to address the drought issue and act on probable mitigation methods, it is vital to detect drought conditions and evaluate their intensity at the early stages. It is very challenging for scientists to predict the drought and its severity with better accuracy. Drought monitoring indices help monitor the drought and measure its severity for a particular location. This chapter discusses in detail various drought indices and their uses. Scientists have developed different indices to measure drought over a specific region

‡ Corresponding Author's Email: nagill_191616@student.nitw.ac.in.

* Corresponding Author's Email: tapasm21@students.ecu.edu.

† Corresponding Author's Email: uday.kumar155@gmail.com.

In: Handbook of Uncertainty in Eurasian Forecasting (HEF)
Editors: Saeid Eslamian, Mir Bintul Huda and Faezeh Eslamian
ISBN: 978-1-68507-966-6

accurately. Recently, various studies used unequal weightage methods to develop composite drought indices. This chapter discusses several techniques for giving weightage to drought indices to build composite drought indices; various drought indices are developed because they are suitable for a particular location. The suitability of that drought index over other regions is conditional based on rainfall, groundwater availability, and several other factors. The value of drought index may increase or decrease based on the factors that we choose Policymakers need to choose a location-specific drought index for effective monitoring of drought.

Introduction

Why is it important to keep an eye on droughts? Water shortages are a naturalistic component of the atmosphere, and they can happen everywhere on the planet, including in deserts and rainforests. Droughts are one of the environmental disasters annually; their consequences are abrupt and harsh, influencing many monetary sectors and people temporally. Droughts have more significant hazards than different threats, generally limited to floodplains, nearshore regions, hurricane paths, or fault regions. Rather than any other threat, droughts lend themselves to observation because of their gradual onset, allowing time to observe progress. Droughts are the most expensive of all environmental hazards, costing the United States six to eight billion dollars annually (Wilhite 2000; Wang 2007; Zhang 2017).

Pervasive and sometimes repeated adverse presence highlights developed and emerging countries to the challenge. Drought is normal, but famine often occurs because of drought. Drought is the primary constraint on farming and animal production across the board. Drought is characterized as a lack of water in the hydrological cycle and can manifest itself in numerous ways. A metric measure of the degree of a drought determines the form and number of the dataset regarding the nature of water scarcity and truncation methods (Oladipo 1985; Masroor 2020).

Drought indices may be considered as either variables or parameters. Examples of weather variables include precipitation, temperature, streamflow, groundwater, river levels, and snowpack (Zargar et al., 2011). Computations of drought intensity, made using climatic and hydrological inputs, include the above mentioned variables. They are simply tracking a single factor, several factors, a hybrid, or a mixture of factors. Defining effective steps for different forms of droughts is incredibly difficult due to the amount and type of consumers and ecosystems impacted. Alternate thresholds and various combinations of inputs generally yield different outputs. "Research should be undertaken before introducing this strategy to define indicators/indices appropriate for pacing, geography, regional environment, and drought. Identifying and quantifying index-dependent decision-making is key to making proper and reliable assessments of drought severity"(Rajsekhar et al. 2015). The weather instruments, that are used to compute various climatic parameters like rainfall, evaporation, transpiration are grouped into the following: (a) Meteorology, (b) Hydrology, (c) Remote sensing, and (d) Composite or aggregated.

Although called 'ease of use,' it is possible that any or none of the indicators will be suitable for implementation by users, depending on their knowledge, resources, the availability of data, and the number of computers. A 'traffic-light' method is applied to each indicator or index according to the following criteria.

The index is considered green if the following conditions are met: the index is easily and publicly available; for each particular situation, regular updates are not necessary; the dataset may contain missing data; the index is already being printed and is online. While an indication of "greenness" may imply that the measure/index applies to any area, it does not mean that it is the best that could be obtained. The selection of indicators and requirements has to be made by the user, and it is highly dependent on the application.

Four of the following factors are important in determining whether an index is yellow: more than one of the same input variables or inputs are required for calculations; an index code or program is not available in the public domain; only a single piece of information is available, but no code is available. Producing the index is simple.

An index is counted as red if it has one of the following properties: an index number would have to be calculated through a code derived from research in the literature; this index is obscure and may be of some use to some, but its popularity is low. The index is part of the complex calculation.

Drought Characterization Concepts

Different drought indices demonstrate many different perceptions of the problem. The word "drought" can be applied to different purposes, ranging from a temporary water shortage to long-term water deprivation. For example, a hydrological drought looks at the effects of insufficient rainfall on hydrology. This section presents the basic concepts of drought that form the basis for developing droughts indices.

Defining drought is in itself a complex thing. In general, people understand drought as a severe rainfall shortage, but objectively, it is challenging to characterize it in terms of planning and management. This is because the context of drought varies from time and place. It can also be defined as a severe water shortage to meet human needs (Zargar et al., 2011). In some cases, drought is not only due to water shortage but also its susceptibility to ground conditions. The complex nature of drought comes from its impacts. Though almost 91 drought impacts have been identified, drought indices use only a handful of impacts like Precipitation, Temperature, Evapotranspiration, Vegetation health, etc.; the conceptual definition of drought is used for policy-making while the operational one (e.g., Hydrological, Agricultural, etc.,) define the beginning, end, and severity of a drought specific to its application.

From an operational perspective, drought can be categorized into three main physical types i.e., (a) Meteorological, (b) Agricultural, and (c) Hydrological droughts (Zargar et al., 2011). In general, these three occur in order. Precipitation deficit causes meteorological drought, which leads to less soil moisture content causing agricultural drought. Low soil moisture leads to higher infiltration of water into the soil causing reduced surface or subsurface runoff resulting lower recharge of water resources like streams and lakes, leading to deferred hydrological drought.

Drought typically manifests itself in four aspects: severity, duration, spatial distribution, and redundancy. Additional characteristics include (are made up of) the following: the frequency of a deficit, predictability, debt growth, occurrence (determine the time when the debt will grow), and timing. Unfortunately, though, the terms severity, intensity, and magnitude have different definitions among cultures. For example, Dracup et al. (1967) used terms such as length, intensity, and duration in Yevich.

The duration of droughts varies from a few weeks to several years. For different timeframes, an area can experience concurrent wet and dry spells. When it is hot or dry in the short term, the area is typically wet in the long term; when it is cold, the region tends to be hot in the short term (NCDC 2010).

Magnitude: It is an accumulated deficit of water with respect to a threshold (e.g., rainfall, soil, or runoff) during a period of drought. It is the positive sum of a drought index value for all the months (Time scale may vary, like if we take it for daily, 16-day, 1-month, 2-months, 3-months, 12-months,) of a drought period.

Intensity: It can be defined as drought magnitude per unit duration.

Severity: The severity of a drought can be defined in two ways: one is the degree of dryness that is generated by the amount of precipitation deficit and the second degree of effects that result from the decrease in precipitation (Wilhite 2004).

Extent: The extent of the drought also varies according to the altitude, region, location, and time of the year. This may be the territory for one or many pixels (cells) and/or areas part of a watershed, or they may be divided into several regions.

Frequency (return period): It can be defined as the average time between two successive drought events of a given severity or even greater than that.

Drought Characterization Using Drought Indices

Apart from precipitation shortage, other variables like evapotranspiration, soil moisture, etc., are also used to characterize drought. Using different models and algorithms such as indicators are used in combination to develop a drought index. These indicators can be meteorological like precipitation, temperature, hydrological ones like stream flow, groundwater level, or agricultural like vegetation and its health.

There are many different methods for quantifying the severity of water shortage or drought; however, the use of drought indices is pervasive (Tsakiris et al., 2007). The single numerical value is obtained by incorporating all the existing measurements of drought indicators into one figure. Many indices can be employed to help with analysis and decision-making when these values are considered together, as opposed to their unadulterated numbers, which are more difficult to analyze (Hayes 2006).

Over the past decades, numerous kinds of drought monitoring indices have been produced. This latest expansion adds further information to our knowledge of the concept of this principle by including further indices that have been developed since the concept was first offered. One of the following serves as the main purpose for using drought indices to define drought characteristics. Drought characterization is specifically important for remediation and policy-making for the drought-affected areas. Drought monitoring indices vary greatly with different drought categories. The main fundamentals of drought monitoring indices vary based on input parameters and their purpose (Niemeyer, 2008; Vicente-Serrano et al., 2010; Zargar et al., 2011).

Development of Drought Indices

Meteorological drought monitoring index: The creation and use of the drought indices are critical to the available data (Steinemann et al. 2005). But these failed to incorporate information from various other variables that became critical only after the 2002 drought (Niemeyer 2008). At the early stages of development, drought indices include only precipitation, such as Rainfall Anamoly Index-RAI (Van Rooy 1965), BMDI-Bhalme-Mooley Drought Index (Bhalme and Petr and others 1980), NRI-National Rainfall Index (Gonzales and Valdés 1994), and EDI-"Effective Drought Index"(González and Wilite 1999). But additional meteorological variables have been considered for the drought indices to have a better correlation and better portray the effects of droughts. This led to a modification of the "Standardized Precipitation Index SPI (McKee et al. 1993) to develop a more comprehensive RDI (Tsakiris et al. 2005)" (Zargar et al., 2011), incorporating transpiration to account for agricultural and hydrological drought effects. The paper by Vicente-Serrano et al. (2010) considered a long-term temperature change. It developed the Standardized Precipitation and Evapotranspiration Index SPEI, whose performance will be similar to SPI in case of absence in long-term temperature trends. KBDI (Keetch-Byram Drought Index) (Keetch and Byram 1968) accounts for temperature and is widely recognized as an important factor in firefighting and monitoring forest fires. On the other hand, PAI-Pálfai Aridity Index (Pálfai 1991) uses groundwater in conjunction with the above parameters and has mainly been used in Hungarian basins.

There are many different ways of identifying an agricultural drought. Agriculture drought monitoring involves soil water balance monitoring and its subsequent deficit during a drought event. This section describes seven indices (non-remote sensing ones) that monitor soil water balance for characterization of agriculture drought: RSM (Relative Soil Moisture) (Narasimhan et al., 1995); Crop Moisture Index "CMI"(Palmer 1968), which is closely related to Palmer Drought Severity Index PDSI, but is useful for only short term drought accounting sub-soil water deficit in the first 5-foot soil layer (Byun et al., 1999; Mey et al., 1993); and CSDI (Crop Specific Drought Index) (Meyer et al. 1993), which is formerly developed for a specific crop that is corn and its variant soya bean. Agricultural Drought Index, DTx (Matera et al. 2007) is used to calculate the deficit in daily transpiration for x days using a soil water balance model named CRITERIA that includes soil, crop, climate conditions, and temperature anomalies that affect evapotranspiration. SMDI and ETDI (Narasimhan and Srinivasan 2005) were developed, seeking higher spatial and temporal resolutions. This was achieved using soil components in the SWAT model with a higher resolution of 16 sq km compared to SPI (7000 sq km) and PDSI (160000 sq km). SMDI (Soil Moisture Deficit Index) uses SWAT to analyze soil profiles up to 2 m from the surface at variable depths like the first 2 feet (SMDI 2), 4 feet (SMDI 4), and 6 feet (SMDI 6). ETDI (Evapotranspiration Deficit Index) and SMDI 2 (which assumes evaporation deficit) are considered for short-term drought monitoring, whereas SMDI 6 is considered for droughts of longer duration. Remote sensing-dependent vegetation indices such as NDVI (Normalised Difference Vegetation Index) (Tucker 1979), EVI (Enhanced Vegetation Index) (Liu and Huete 1995, Liu and Kogan 1995), and Temperature Condition Index -TCI (Gao 1996) are also employed to assess the overall health of the vegetation of a region (Sivakumar et al. 2011).

"Hydrological Drought Indices are designed to capture the slow impacts on hydrology impacts of drought. Before, the sophisticated PHDI (Palmer Hydrological Drought Index) (Palmer 1965)"((Zargar et al., 2011) model considering the variables rainfall, recharge, soil water content, and the losses (also known as evapotranspiration, evaporation, and water runoff) was used. One of the most popular PDSI families of indices, SWSI, is developed from the PDSI's lack of snow accumulation component (Shafer and Dezman 1982). "Later, RDI (Reclamation Drought Index) (Weghorst 1996) improved SWSI (Surface Water Supply Index) by incorporating the variability of water demand" (Zargar et al., 2011).

Major Operational Drought Indices

In this, we have covered six indices of drought that are often employed in all the areas. They are commonly found and therefore deserve more detailing than others. Few indices are focused on a single type of impact or application, while others are capable of reacting to various effects and thus are used to monitor various types of drought. An important example is SPI, which can be applied for shorter and longer time scales to consider both hydrological and agricultural droughts and their impacts. Standardized Precipitation Index, in short, SPI (McKee et al. 1993), is a precipitation-only based meteorological drought index. It compares precipitation with its multi-year average value. SPI overcomes the discrepancy in the recorded precipitation by transforming them into normally distributed data. The data points are fit to a gamma distribution at first, and then they are converted into normal distribution using an equal-probability transformation (Rajsekhar et al. 2015, Zargar et al., 2011). When the mean is set to zero, above the zero indicates wet periods, and below the zero indicates dry ones. SPI value represents the deviation of cumulative precipitation deficit from its normalized average by how many standard deviations it is away from (Drought Watch 2010). If the value drops to 0 or less and is consistent at the same level for a period of time, then SPI reaches a value of -1 or below, which is when we say the drought has occurred (McKee et al. 1993). An important component of SPI is its ability to calculate the drought levels at varied time scales. It can be computed for any period; however, it is usually applied for 3, 6, 12, and 48 months. As time passes, deficits may have different effects on various features (for example, streams, groundwater, or snow) that can be detected using the multitude of a feature of SPI. The Palmer Drought Severity Index is a widely known meteorological index mostly used in the United States. The PDSI (Palmer 1965) bases its drought concepts on water supply and demand information. Emphasis is placed on the lack of moisture rather than weather events that deviate from normal. PDSI uses precipitation, temperature, and local data on moisture content to predict soil conditions. Using the mentioned inputs, PDSI can compute four variables in the water balance equation viz., evapotranspiration, runoff, soil moisture, and gain in moisture from evapotranspiration (Zargar et al., 2011).

The latest report issued by the United States Drought Monitor(USDM) shows that the likelihood of being affected by drought in the USA has increased significantly, as more than half of the nation has been afflicted by drought since 1936. The USDM (Svoboda et al. 2002) consists of a composite index incorporating soil moisture, drought, and precipitation (Zargar et al., 2011). The USDM combines different indices, like SPI and PDSI, and others like soil moisture and precipitation into a drought map each week. Experts afterward process this information for fine-tuning. Due to its diverse nature, USDM is available to meet various needs of water planners and the agricultural industry. In theory, it is widely used for research, but not in practice or by the media. The index is getting more pervasive and popular outside of the United States.

NDVI is used to measure vegetation conditions and can be estimated using remote sensing indices (Rouse et al., 1974). The Advanced High-resolution Red Radiometer (AVHRR) sensors reflect red and near-infrared radiations, which are used in the case of NDVI. Near-infrared reflectance is equivalent to the visible red reflectance under normal or favorable conditions. Chlorophyll (the substance that plants use to generate carbohydrates) reflects less R(red) by absorbing light when plants are in healthy condition, and thereby low R values yield higher NDVI. Plants have higher R as well as a lower NDVI when they are unhealthy. NDVI has been used as the basis for several other remote sensing indices, such as the Vegetation Condition Index (VCI)(Kogan 1990) and Vegetation Health Index (VHI).

Table 1. Advantages and disadvantages of popular drought indices

Drought Index,	Pros	Cons
SPI (McKee et al. 1993) Precipitation	Simplicity; SPI relies only on precipitation data. As SPI is adaptable for analyzing drought at variable time scales, it can be used for monitoring agricultural and hydrological droughts. Comparing precipitation departure from normal for various regions with highly different climates is possible. It equally represents both wet and dry climates and can be used to monitor wet periods.	Uses only precipitation, loosely connected to ground conditions. Potential evapotranspiration is a valuable additional indicator
PDSI (Palmer 1965) Precipitation, temperature	PDSI monitors damp spans more thoroughly than only precipitation-based drought monitoring index.	More sophisticated algorithms and calibration are required for using PDSI for regions close to the mountain.
NDVI (multiple)Visible red bands, near-infrared bands	Easy algorithm and actual measure of drought without extrapolating the observations.	Saturation of NDVI value over the region close to dense vegetation and atmospheric interference are major limitations of the NDVI.
"SPEI" (Vicente-Serrano et al. 2010)	It is similar to SPI. It overcomes the defect of SPI being unable to identify the pattern of increase in drought duration and magnitude resulting from higher temperatures.	Where there are no apparent changes in temperature with time, it is almost equal to SPI or other precipitation-based drought indices.
"Vegetation Condition Index (VCI)" Kogan (1990	It is based on NDVI. It measures the health of the vegetation, which gives us the scenario of drought in a region.	As it is based on NDVI, which in turn is a remote sensing-based index, there is a chance for cloud contamination in the data obtained.
Crop Moisture Index (CMI) Palmer (1968)	It is calculated in the same way as PDSI. It gives us the weighted output that is useful in comparing two different regions. It's quick to respond to a rapid change in climatic conditions.	It was originally developed for grain-producing regions in the U.S. It may show false recovery from droughts, but short-term measures may not be enough to recover from long-term droughts.

The Palmer Crop Moisture Index, or CMI, was developed by Palmer (1968) from the PDSI calculation procedures that were already established within the calculation of regions that produce major crops. It is not used to take long-term droughts into account, but one of its advantages is that it identifies dry spells susceptible to agricultural areas. Weekly crop conditions are monitored using a meteorological approach using this index. The difference between the two indices is that. In contrast, the PDSI looks at long-term moisture conditions in the climate zones; the CMI was built to examine regional crop-specific moisture conditions. This is calculated on the average temperature and total precipitation of each week during the climate division and the CMI of the previous week. The CMI reacts quickly and is influenced by location and time; thus, weekly CMI maps are commonly utilized to compare moisture conditions across the country over different regions. The index is not suitable for long-term crop monitoring because it is designed to predict drought, which is typically short-term. In addition, index reactions happen quickly for the short term but tend to fail to reflect long-term drought trends. For example, beneficial rainfall may signify adequate moisture for the CMI value while it remains in place, but in the long term, drought may still persist. Another shortcoming of the CMI is that it is unemployed as a long-term drought monitor because the index starts and finishes each growing season at zero.

This index, abbreviated as SWSI developed by Shafer & Dezman (1982), utilizes streamflow, precipitation, and reservoir data as its input. This value represents the basin's conditions and has the disadvantage of having to be recalculated whenever there is a data station change or management of water supply, which limits inter-basin comparisons. Since the SWSI is intended for relatively homogeneous regions, it was not made for topographic variations and did not account for snow and subsequent runoff. The index calculates both hydrological and climatic parameters into a single value that closely resembles the Palmer Index. It is possible to standardize the index values to allow comparisons between basins.

Like the SWSI, the Reclamation Drought Index calculates its output at the river basin, which uses temperature, precipitation, flow, and reservoir data and level as input, which gives it the benefit of also accounting for evaporation. The Bureau of Reclamation U.S. developed the index as a means to trigger money from emergency drought relief sources and has been used to determine drought severity and duration.

Expanding the Remote-Sensing Capacity

Many indices use R.S. (Remote Sensing) data, with new ones constantly in the proposal. While indices like NDVI, NDWI are few of the widely known remote sensing indices, there are VCI, TCI, that are derived from those core NDVI, land surface temperature etc., are being constantly proposed to overcome the problems to their predecessors in effective drought mnoitiring. Researchers have made new sensors and algorithms that continually enable the integration of remote sensing information into the characterization of drought. Modern sensors have a high spatial resolution, but that problem still exists with drought indices products (Niemeyer 2008). New thematic enhancement algorithms and other environmental enhancement algorithms improve data reduction algorithms and the accuracy of remote sensing-based products (Rajsekhar et al. (2015)). As state-of-the-art in remote sensing improves, new and improved indices are introduced. One currently popular one is NDVI, and other indices like VegDRI (Vegetation Drought Response Index), VCI (Kogan 1990), TCI, and VHI (Kogan 1995) are in operation (NDMC 2011; NOAA 2011). They are traditionally employed to detect near-infrared (NIR) and short-wavelength infrared (SWIR), and Red bands that are more susceptible to change concerning drought variables (Zargar et al., 2011). Land Surface Temperature (LST) was found to be more appropriate when used in combination with NDVI (greenness) for drought prediction (Cai et al. 2011; Lambin and Ehrlich 1995; Prihodko and Goward 1997; Rhee et al. 2010; Wan et al. 2004; Wang et al. 2001) which can also be determined using remotely sensed satellite data. Variables like precipitation and soil moisture, which are more reactive toward a drought, are widely used for drought monitoring. But we need the data to be continuous temporally and spatially, which is impossible in conventional data acquisition methods.

In such cases, we can adopt remote sensing variables like TRMM rainfall or AVHRR rainfall GLDAS-based soil moisture for drought monitoring. Though not accurate as conventional data, these variables can provide an insight into the temporal and spatial variation of the data, which is primary in drought monitoring and assessment. Satellites like LANDSAT-8, newly launched LANDSAT-9, Sentinel-2, etc., have a higher resolution as compared to their previous versions and are more accurate in representing the land conditions as they are during all seasons. Bands of the images from those satellites are used to calculate NDVI, NDWI, VCI, LST, and other similar indices, which are widely used in many parts of the world for drought detection and monitoring. Images from these new satellites are of higher resolution, so they occupy more disk space when downloaded, reducing the available disk space. It is also difficult to perform the analysis with such high-resolution data on a pc iht a normal configuration. This can be overcome by Google Earth Engine (GEE), which processes data at a higher speed. GEE is an online platform where you can access the remote sensing data without downloading it, run the analysis, and download the final results. We can access the data via a simple code of JavaScript or Python. There is a manual to learn more about the keywords and symbols that can be used in GEE. More additions and changes are coming up in GEE. Still, platforms like GEE make remote sensing data easily accessible to researchers and engineers that can be used for planning, development, decision, and policy making. Before that, a thorough survey of remote-sensing drought indices can be informative and useful for further analysis and development of drought characterization using remote sensing data.

Multivariate Drought Index

Among the existing drought indices, the ones that characterize drought using a single variable are the most common. Even among highly operational ones, indices with a single variable or combination of two variables (e.g., precipitation and temperature or precipitation and streamflow) are highly customary. They may not be precise enough to detect the current state of the drought. These consider only one specific physical form of drought: hydrological, meteorological, or agricultural. This study might not be able to get a complete overview of the drought, as it is subject to many variables. But, from this information, it follows that different drought indices would generally disagree with one another (Zargar et al., 2011). A specific area may be experiencing only one type of drought, but the multiple varieties of drought may be widespread elsewhere. It is always better to comprehensively represent all physical drought variables, such as meteorological, hydrological, and agricultural droughts. MDI may include precipitation, runoff, evapotranspiration, and soil moisture measurements, thus representing all the physical varieties of drought overcoming the limitations by considering more than two variables (Rajsekhar et al., 2015).

Non-hybrid drought indices that do not utilize hybrid descriptors are mainly useful for a specific location, time, and objective, not providing any comprehensive information about a drought event. Integrating various drought indices have been given greater consideration recently, particularly due to their successful incorporation and demonstrated ability to work with the available, field-based indicators (Kallis 2008; Niemeyer 2008; Sivakumar et al. 2011). Karamouz et al. (2009) integrated SPI, SWSI, and PDSI to develop the HDI. In response to the non-stationarity in predicted future climate scenarios (IPCC 2007) which may adversely affect different aspects of Earth's climate variables, research to include temporal variations of the future climate is also being initiated. The SPEI attributes the prolonged and more pronounced droughts to rising temperatures (Vicente-Serrano et al. 2010).

Objective Blend Drought Index (OBDI), developed by Svoboda et al. (2002), is a composite drought index that considers a weighted average of all physical forms of drought through variables like precipitation, evaporation, reservoir storage, and soil moisture. Aggregated Drought Index (ADI; Keyantash and Dracup, 2004), similar to OBDI, consider variables accounting for various forms of drought by integrating them into a single time series using Principal Component Analysis (PCA).

Mutlivariate Integrated Drought Monitoring Index (MIDMI) (Mahesh et al., 2022) is a remote sensing based agricultural drought monitoring index which is developed by comparison of various other remote sensing index like NDVI, NDWI etc., using unsupervised classification and percent area under each drought category. MDMI uses the five different drought indices to combine them by assigning unequal weights to each of those five indices using Iyengar Sudarshan method (Iyengar & Sudarshan 1982).

Summary and Conclusion

Multiple drought types use different variables that may or may not affect a place simultaneously (Zargar et al., 2011). So, we need a multivariate drought monitoring index that jointly represents various drought forms a region can experience at different periods.

The ranks generated from the various combinations of weighting and aggregation show variation among the countries. The best policy in developing the aggregated drought index is to study the drought indices best suited for a particular area according to particular climatic conditions. To develop such multivariate drought monitoring indices, data has to be continuous, which is impossible with conventional data acquisition methods. Especially in underdeveloped countries, data obtained from conventional acquisition methods are very scarce to use in developing multivariate indices. Only some important variables like rainfall and temperature are easily available, whereas other variables like vegetation, evapotranspiration, land use and land cover changes, etc. Which also affects drought and its severity to much extent. One solution to this problem is to use remote sensing data.

Several new droughts monitoring indices use remote sensing-based data, allowing for continuous land monitoring (Aghakouchak, 2015; Cardona, 2005; Huang et al., 2016). The usage of remote sensing data to monitor drought and its conditions became a general procedure since Kogan's development of the Vegetation Condition Index (VCI) (1995a,b).

By transforming the data obtained for the variables under study into pure, dimensionless numbers, the data is normalized to bring the indicators to the same standard. Four weighting techniques are used to weight the normalized variables: the Iyengar-Sudarshan Method, the Benefit of Doubt Approach, Principal Component Analysis, and the Unobserved Component Model (Bhandari et al., 2012; Duong et al., 2017; Das et al., 2021).

References

Aghakouchak, A. (2015). A multivariate approach for persistence-based drought prediction: Application to the 2010–2011 East Africa drought. *Journal of Hydrology*, *526*, 127–135. DOI: 10.1016/j.jhydrol.2014.09.063.

Alley, W. M. (1984). The Palmer Drought Severity Index: Limitations and assumptions. *Journal of Climate and Applied Meteorology,* 23(7), 1100–1109.

Al-Najjar, H., Ceribasi, G., Dogan, E., Abualtayef, M., Qahman, K., &Shaqfa, A. (2020). Stochastic time-series models for drought assessment in the Gaza Strip (Palestine). *Journal of Water and Climate Change,* 11(S1), 85–114. https://doi.org/10.2166/wcc. 2020. 330.

Bhandari, A. K., Kumar, A., & Singh, G. K. (2012). Feature extraction using Normalized Difference Vegetation Index (NDVI): A case study of Jabalpur City. *Procedia Technology,* 6, 612–621. https://doi.org/10.1016/j.protcy.2012.10.074.

Bhangale, U., More, S., Shaikh, T., Patil, S., & More, N. (2020). Analysis of surface water resources using sentinel-2 imagery. *Procedia Computer Science,* 171, 2645–2654. https://doi.org/10.1016/j.procs.2020.04.287.

Bhatt, D., & Mall, R. (2015). Surface water resources, climate change and simulation modeling. *Aquatic Procedia*, *4*, 730–738. https://doi.org/10.1016/j.aqpro.2015.02.09 4.

Byun, H.-R., & Wilhite, D. A. (1999). Objective quantification of drought severity and duration. *Journal of Climate,* 12(9), 2747–2756.

Cardona O. D. (2005). *Indicators for Disaster Risk and Risk Management.* Program for Latin America and the Caribbean: Summary Report. Instituto de Estudios Ambientales. Universidad Nacional de Columbia, Manizales, Columbia.

Cardona O. D. van Aalst M. K. Birkmann J. Fordham M. McGregor G. Perez R. Pulwarty R. S. Schipper E. L. F. Sinh B. T. (2012). Determinants of risk: exposure and vulnerability. In: *Managing the Risks of Extreme Events and Disasters to Advance Climate Change Adaptation. A Special Report of Working Groups I and II of the Intergovernmental Panel on Climate Change IPCC* (Field C. B. Barros V. Stocker T. F. Qin D. Dokken D. J. Ebi K. L. Mastrandrea M. D. Mach K. J. Plattner G.-K. Allen S. K. Tignor M. Midgley M. , eds). Cambridge University Press, Cambridge, UK and New York, USA.

Das, S., Das, J., & Umamahesh, N. V. (2021). Nonstationary modeling of meteorological droughts: application to a region in India. *Journal of Hydrologic Engineering,* 26(2), 05020048. https://doi.org/10.1061/(asce)he.1943-5584.0002039.

Dracup, J. A., Lee, K. S., & Paulson, E. G., Jr. (1980). On the definition of droughts. *Water Resources Research,* 16(2), 297–302.

Duong, H. H., Thuc, T., &Ribbe, L. (2017). Assessing and calculating a climate change vulnerability index for agriculture production in the Red River Delta, Vietnam. In *Redefining Diversity & Dynamics of Natural Resources Management in Asia,* 2, 27–40. https://doi.org/10.1016/b978-0-12-805453-6.00003-6.

Gao, B. (1996). NDWI—A normalized difference water index for remote sensing of vegetation liquid water from space. *Remote Sensing of Environment,* 58(3), 257–266. https://doi.org/10.1016/s0034-4257(96)00067-3.

Guttman, N. B. (1999). Accepting the standardized precipitation index: A calculation algorithm. *Journal of the American Water Resources Association,* 35(2), 311–322. https://doi.org/10.1111/j.1752-1688.1999.tb03592.x.

Hao, Z., & AghaKouchak, A. (2013). Multivariate Standardized Drought Index: A parametric multi-index model. *Advances in Water Resources,* 57, 12–18. https://doi.org/10.1016/j.advwatres.2013.03.009.

Hao, Z., & Aghakouchak, A. (2014). A nonparametric multivariate multi-index drought monitoring framework. *Journal of Hydrometeorology,* 15(1), 89–101. https://doi.org/10.1175/JHM-D-12-0160.1.

Heim R. R. Jr . (2002). A review of twentieth-century drought indices used in the United States. *Bulletin of the American Meteorological Society* 83 (8), 1149–1166.

Hong, Z., Zhang, W., Yu, C., Zhang, D., Li, L., & Meng, L. (2018). SWCTI: Surface Water Content Temperature Index for Assessment of Surface Soil Moisture Status. *Sensors,* 18(9), 2875. https://doi.org/10.3390/s18092875.

Huang, S., Huang, Q., Leng, G., & Liu, S. (2016). A nonparametric multivariate standardized drought index for characterizing socioeconomic drought: A case study in the Heihe River Basin. *Journal of Hydrology, 542*, 875–883. https://doi.org/10.1016/j.jhydrol.2016.09.059.

Iyengar, N. S. and Sudarshan. P. (1982). A method of classifying regions multivariate data. *Economic and Political Weekly,* Special Article: 2048-52.

Jagadish Padhiary, Kanhu Charan Patra, Sonam Sandeep Dash, A. Uday Kumar; Climate change impact assessment on hydrological fluxes based on ensemble GCM outputs: a case study in eastern Indian River Basin. *Journal of Water and Climate Change*, 11 (4): 1676–1694. doi: https://doi.org/10.2166/wcc.2019.080.

Jiao, W., Wang, L., Novick, K. A., & Chang, Q. (2019). A new station-enabled multi-sensor integrated index for drought monitoring. *Journal of Hydrology*, *574*, 169–180.

Kao, S.-C., &Govindaraju, R. S. (2010). A copula-based joint deficit index for droughts. *Journal of Hydrology,* 380(1–2), 121–134. https://doi.org/10.1016/j.jhydrol.2009.10.029.

Katipoğlu, O. M., Acar, R., &Şengül, S. (2020). Comparison of meteorological indices for drought monitoring and evaluating: a case study from Euphrates basin, Turkey. *Journal of Water and Climate Change,* 11(S1), 29–43. https://doi.org/10.2166/wcc.2020.171.

Keyantash, J. A., &Dracup, J. A. (2004). An aggregate drought index: Assessing drought severity based on fluctuations in the hydrologic cycle and surface water storage. *Water Resources Research,* 40(9). https://doi.org/10.1029/2003wr002610.

Kogan, F. N. (1990). Remote sensing of weather impacts on vegetation in non-homogeneous areas. *International Journal of Remote Sensing,* 11(8), 1405–1419. https://doi.org/10.1080/01431169008955102.

Kogan, F. N. (1995). Application of vegetation index and brightness temperature for drought detection. *Advances in Space Research,* 15(11), 91–100. https://doi.org/10.1016/0273-1177(95)00079-T.

Kogan, F. N. (1995a). Drought of the late 1980s in the United States as derived from NOAA polar-orbiting satellite data. *Bulletin of the American Meteorological Society*, 76, 655−668.

Kogan, F. N. (1995b). Application of vegetation index and brightness temperature for drought detection. *Advanced Space Research,* 15, 91−100.

Kogan, F.N. (2000). Contribution of remote sensing to drought early warning. In Early Warning Systems for Drought Preparedness and Drought Management, *Proceedings of an expert group meeting held on Warning Systems for Drought Preparedness and Drought Management.* Edited by D.A. Wilhite, M.V.K. Sivakumar, and D.A. Wood. Lisbon, Portugal. 75–87.

Kogan, F. N. (2001). Operational space technology for global vegetation assessment. *Bulletin of the American Meteorological Society,* 82(9), 1949–1964.

Limones, N., Marzo-Artigas, J., Wijnen, M., &Serrat-Capdevila, A. (2020). Evaluating drought risk in data-scarce contexts. The case of southern Angola. *Journal of Water and Climate Change,* 11(S1), 44–67. https://doi.org/10.2166/wcc.2020.101

Masroor, M., Rehman, S., Avtar, R., Sahana, M., Ahmed, R., & Sajjad, H. (2020). Exploring climate variability and its impact on drought occurrence: Evidence from Godavari Middle sub-basin, India. *Weather and Climate Extremes,* 30, 100277. https://doi.org/10.1016/j.wace.2020.100277.

Matera, A., Fontana, G., and Marletto, V. 2007. Use of a new agricultural drought index within a regional drought observatory. In *Methods and Tools for Drought Analysis and Management*. Edited by G. Rossi. Springer, Dordrecht, The Netherlands. pp. 103–124

McFeeters, S. K. (1996). The use of the Normalized Difference Water Index (NDWI) in the delineation of open water features. *International Journal of Remote Sensing,* 17(7), 1425–1432. https://doi.org/10.1080/01431169608948714

McKee, T. B., Kleist, Doesken, N. J., John, Science, The relationship of drought frequency and duration to time scales. In *Proceedings of the 8th Conference on Applied Climatology,* Anaheim, Calif. 17– 22 January 1993. American Meteorological Society.

Meyer, S.J., Hubbard, K.G., and Wilhite, D.A. 1993. A crop-specific drought index for corn. I: Model development and validation. Agron. J. 85 (2): 388–395.

Narasimhan, B., and Srinivasan, R. 2005. Development and evaluation of Soil Moisture Deficit Index (SMDI) and Evapotranspiration Deficit Index (ETDI) for agricultural drought monitoring. *Agric. For. Meteorol*. 133 (1–4): 69–88. doi: 10.1016/j.agrformet.2005.07.012

Olukayode Oladipo, E. (1985). A comparative performance analysis of three meteorological drought indices. *Journal of Climatology,* 5(6), 655–664. https://doi.org/10.1002/joc.3370050607

Padhiary, J, K C Patra, S. S. Dash, and A. Uday Kumar. Climate change impact assessment on hydrological fluxes based on ensemble GCM outputs: a case study in eastern Indian River Basin. *Journal of Water and Climate Change*, 11(4): 1676-1694.

Palmer, W.C. (1965). Meteorological drought. *Weather Bureau Research Paper No. 45*, US Deptartment of Commerce, Washington, DC. 58 pp.

Palmer, W. C. (1968). *Keeping Track of Crop Moisture Conditions,* Nationwide: The New Crop Moisture Index. Weatherwise, 21(4), 156–161.

Rajsekhar, D., Singh, V. P., & Mishra, A. K. (2015). Multivariate drought index: An information theory based approach for integrated drought assessment. *Journal of Hydrology,* 526, 164-182.

Rajsekhar, Deepthi & Mishra, Ashok. (2014). Multivariate drought index: An information theory-based approach for integrated drought assessment. *Journal of Hydrology*. 526. doi: 10.1016/j.jhydrol.2014.11.031.

Shafer, B.A. &Dezman, L.E. (1982), Development of a surface water supply index to assess the severity of drought conditions in snow-pack runoff areas. *Proceedings of the Western Snow Conference,* 164-175.

Sivakumar, M. V. K., Motha, R. P., Wilhite, D. A., and Wood, D. A. 2011. *Agricultural Drought Indices — Proceedings of an Expert Meeting*. 219 pp. Available from: http://www.whycos.org/WMO/clw/agm/documents/AgriculturalDroughtIndicesProceedings29311.pdf.

Springs, C., Carolina, N., State, I., Survey, W., Svoboda, M., & Drought, N. (2002). *Drought Monitor.* April, 1181–1190.

USGS.gov. n.d. NDVI, the Foundation for Remote Sensing Phenology https://www.usgs.gov/core-science-systems/eros/phenology/science/ndvi-foundation-remote-sensing-phenology?qt-science_center_objects=0#qt-science_center_objects (Accessed the 28th of February 2021).

Wang, L., & Qu, J. J. (2007). NMDI: A normalized multi-band drought index for monitoring soil and vegetation moisture with satellite remote sensing. *Geophysical Research Letters,* 34(20). https://doi.org/10.1029/2007gl031021.

Welikhe, P., Quansah, J. E., Fall, S., &McElhenney, W. (2017). Estimation of soil moisture percentage using LANDSAT-based moisture stress index. *Journal of Remote Sensing & GIS,* 06(02). https://doi.org/10.4172/2469-4134.1000200.

Wilhite, D. A. (2000). Drought as a natural hazard: Concepts and definitions. *Drought: A Global Assessment,* 3–18.

Wilhite, D.A. (2004). Drought as a natural hazard. In *International Perspectives on Natural Disasters: Occurrence, Mitigation, and Consequences.* Edited by J.P. Stoltman, J. Lidstone, and L.M echano. Kluwer Academic Publishers, Dordrecht, The Netherlands. 147–162.

Yevjevich, Vujica M. (1969). An objective approach to definitions and investigations of continental hydrologic droughts. *Journal of Hydrology,* 7(3), 353. https://doi.org/ 10.1016/0022-1694(69) 90110-3

Zargar, A., Sadiq, R., Naser, B., & Khan, F. I. (2011). A review of drought indices. *Environmental Reviews,* 19(1), 333–349.

Zhang, X., Chen, N., Li, J., Chen, Z., & Niyogi, D. (2017). Multi-sensor integrated framework and index for agricultural drought monitoring. *Remote Sensing of Environment*, *188*, 141–163.

Chapter 5

Flood Forecasting and Disaster Risk Management: A Case Study of the Danube River

Balatonyi Laszlo*

EU Strategy for the Danube Region,
Environmental Risks Priority Area,
Vienna, Austria

Abstract

The Danube and its tributaries have been crossing mountains and plains in their almost unchanged riverbeds for thousands of years, disregarding national and administrative boundaries. Nevertheless, even decades ago, several countries made only limited data and information available on the water level and flood protection status of their rivers. In recent years, information has changed hands based on bilateral agreements, but for others, information could only be obtained by browsing the Internet, sometimes in a rather complicated way. EU Strategy for the Danube Region Environmental Risks Priority Area initiated a project that is supporting ICPDR in developing the Danube Hydrological Information System. The main aim of the Danube River Basin Enhanced Flood Forecasting Cooperation (DAREFFORT) project is to give a comprehensive overview of the complex national flood and ice forecasting systems and eliminate the shortcomings of the existing forecasting practices as well improve the exchange and availability of hydrological and meteorological data between the participating countries with establishment of the Danube Hydrological Information System. The Evaluation report on flood and ice forecasting systems and

* Corresponding Author's Email: balatonyi.laszlo@ovf.hu.

In: Handbook of Uncertainty in Eurasian Forecasting (HEF)
Editors: Saeid Eslamian, Mir Bintul Huda and Faezeh Eslamian
ISBN: 978-1-68507-966-6

methodologies in the Danube countries produced in the frame of the DAREFFORT project provides more detailed information on flood information services in DRB. In the frame of the DAREFFORT project, all authorized meteorological and hydrological data of the Danube River are stored in a central database and made available online to all licensed hydrological and flood protection institutions for further processing in virtually real-time. At this moment 12 countries of the Danube have joined forces to work out their proposals that are essential for the future.

Keywords: Danube river basin, Danube, Hydrological Information System, flood forecasting, ice forecasting systems, forecasting systems, macro-regional strategy, environmental risks, priority area

Introduction

The Danube River Basin (hereafter: DRB) is the "most international" river basin in the world covering territories of 19 countries. Those 14 countries with territories greater than 2,000 km^2 in the DRB cooperate in the framework of the International Commission for the Protection of the Danube River (hereafter: ICPDR) the DRB is the second largest in Europe with an area of 807,827 km^2. In accordance with the EU Water Framework Directives, the Danube and its tributaries, transitional waters, lakes, coastal waters, and groundwater form the Danube River Basin District (hereafter: DRBD). The DRBD covers the DRB, the Black Sea coastal catchments in Romanian territory, and the Black Sea coastal waters along the Romanian and partly Ukrainian coasts.

The work of the ICPDR is based on the Danube River Protection Convention (DRPC), the major legal instrument for cooperation and transboundary water management in the Danube River Basin.

EU Strategy for the Danube

Floods, drought, wildfires, and low-flow events, as well as water scarcity situations and extreme storms, are likely to become more intense, longer, and more frequent. Natural disasters are perceived as a worldwide increasing phenomenon which seriously impact human life (Natural disasters 2019). Van Aalst (2006) proves that there are several reasons for this, however, the key one is climate change which results in severe atmospheric and hydrologic events. An intensification of floods, hurricanes, droughts, and other disasters obviously leads to high losses in human life. Furthermore, it has devastating consequences for private and public property, agriculture, forestry, industry, and critical infrastructure having a specific feature of multiplying negative impact through cascading effects (Lawrence et al., 2020, Setola Theocharidou 2017, Zwęgliński et al., 2020). An increase in air and water temperature, combined with changes in precipitation, water availability, water quality, and the increase in extreme events may lead to changes in ecosystems, life cycles, and biodiversity in the Disaster Resistance Business (DRB) in the long-term. This is frequently mentioned to be one of the most relevant climate change consequences (ICPDR 2019). Therefore, high impact, low probability events, such as major disasters, are recently more and more common. Among them, flood is broadly recognized to be the highest risk event among natural disasters in Europe and worldwide.

Globally, in 2019 this type of disaster took the lives of more than five thousand people and affected 31 million others. In recent years this anxious tendency is also observed on a regional level. For example, in the Danube River Basin, most notably in 2002, 2006, 2013, and 2014, parts of the district were affected by very strong or extreme flooding events. They caused a significant human and economic impact. In 2006, four casualties were reported in the Czech Republic and Slovakia. The costs of damage amounted to almost 600 million euros in the whole basin. In 2010, it increased even further and therefore, there were 35 casualties, and financial damage amounting to around two billion euros. The figure was surpassed to a greater extent in 2013 with 2.3 billion euros of damage, mostly in Germany and Austria as well as nine additional casualties reported from Austria and Romania. And, most recently, the Sava River Basin in Croatia, Bosnia, and Herzegovina as well as Serbia was hit very hard in May 2014, affecting 2.6 million people, killing 79, and causing almost four billion Euro of further damage across the three countries. As the impacts of climate change are getting more drastic, the situation is getting more complicated. The three biggest floods on the Danube happened in the last two decades, whereas on the Tisza four record-breaking floods occurred within 36 months around the millennium. The floods of the past 20 years made it clear that Hungary must be prepared for extreme floods in the future. Flood forecasting, data collection, and harmonized data sharing are becoming importance along with the countries of the Danube River Basin. Precipitation, water stage, discharge, water temperature, and ice phenomena are measured isolated in each country.

The EU Strategy for the Danube Region (hereafter: EUSDR) is one of four EU macro-regional strategies (MRS) in Europe. The EUSDR provides an integrated framework for strengthening this cooperation between nations. Bringing together 115 million people from nine EU member states, three EU candidate countries, and two EU neighbor countries, it has an important integrative and cohesive function.

The Danube Region Strategy addresses a wide range of issues; these are divided into 4 pillars and 12 priority areas. Each priority area is managed by two countries as Priority Area Coordinators. The coordination of the Environmental Risks Priority Area (hereafter: EUSDR PA5) is managed by Hungary and Romania. The main focus of the work is to address the challenges of water scarcity and droughts in line with the Danube River Basin Management Plan, the report on the impacts of droughts in the Danube Basin and the ongoing work in the field of climate adaptation. Therefore, in the past few years, EUSDR PA5 contributed to the elaboration of the ICPDR Climate Change Adaptation Strategy Update 2018, supported project elaboration and implementation in the field of drought management and climate change-related spatial planning, disseminate scientific results to anticipate regional and local impacts of climate change through research.

Flood risk management is also a significant target of the priority area. To achieve a reduction of flood risk events, EUSDR PA5 provides and enhances continuous support to the implementation of the Danube Flood Risk Management Plan. In case these prevention measures are not effective enough, then disasters occur, therefore EUSDR PA5 supports the assessment of disaster risks in the Danube Region, encouraging actions to promote disaster resilience, preparedness, and response activities.

Flood Forecasting in EU, European Flood Awareness System

After the Danube and Elbe floods in 2002, the European Commission initiated the development of a European Flood Awareness System (EFAS) to increase the preparedness for floods in Europe. EFAS was developed in close collaboration with the ICPDR and the national hydro-meteorological services sharing the Danube river basin amongst others. The aim of EFAS is to gain time for preparedness measures before major flood events strike, particularly for large trans-national river basins such as the Danube, both on the country as well as the European level. This is achieved by providing complementary, added value information to the national hydrological services and by keeping the European Response and Coordination Centre[1] informed about ongoing floods and about the possibility of upcoming floods across Europe. Since 2012 EFAS is running fully operational as part of the Copernicus Emergency Management Service[2].

EFAS provides the national authorities with a wide range of flood forecast information such as medium-range flood forecasts with a lead time of 10-15 days, impact forecasts, flash flood forecasts with up to 48 hours lead time, and seasonal outlooks for the coming months. The information can be accessed through web services. All relevant flood forecasting authorities in the Danube river basin are EFAS partners. EFAS is constantly being further developed with regular new model calibrations including more in-situ data or changes to the model setup and improving its products. Through collaboration at the Danube river basin as well as at the European scale EFAS fosters knowledge exchange and data sharing amongst the national hydro-meteorological authorities and hence is an essential tool to improve overall flood risk management in the Danube river basin.

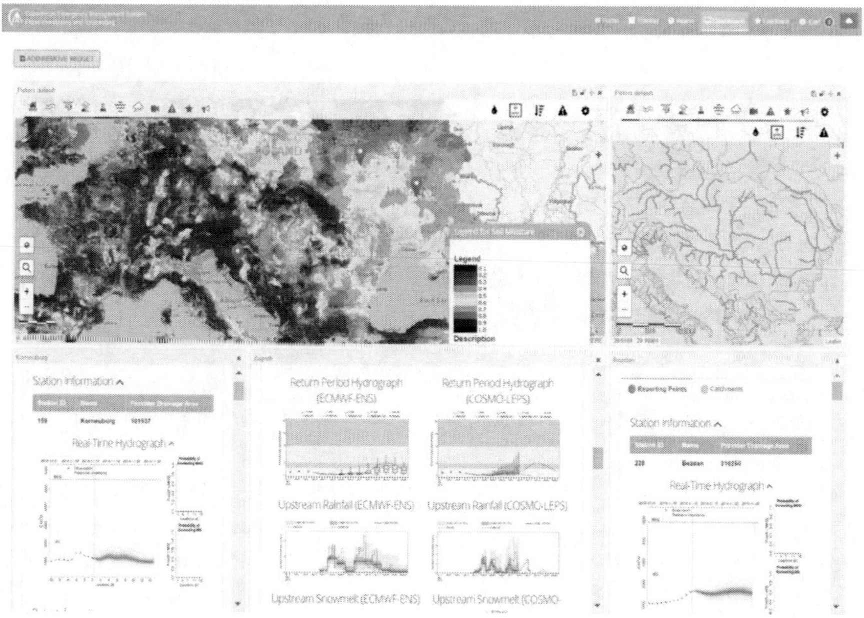

Figure 1. Screenshot of the EFAS web interface.

[1] http://ec.europa.eu/echo/en/what/civil-protection/emergency-response-coordination-centre-ercc.

[2] https://emergency.copernicus.eu/.

Danube River Basin Enhanced Flood Forecasting Cooperation

The reliable and comprehensive hydrologic data is the basis for sound forecasting in any country of the Danube Basin. None of the flood risk mitigation measures serves better for the protection of human lives and the social estate than enhancing the preparation time to avoid catastrophes. The most cost-effective non-structural tangible solution which highly reflects the solidarity principle is the improvement of forecasting capabilities on a basin-wide scale.

The Danube River Basin Enhanced Flood Forecasting Cooperation (hereafter: DAREFFORT) project delivers an outstanding overview of the present status of the national forecasting capabilities from the partners and the stakeholders from the DRB. In the frame of the DAREFFORT project the partners jointly worked out the policy recommendations for the ICPDR in support of the establishment of the Danube Hydrological Information System (hereafter: DanubeHIS) which is a fundamental step toward flexible and sustainable data exchange for the DRB countries. The main focus is to enhance access to the recorded hydrological data and to provide harmonized distribution for all the countries in the Danube catchment. For this purpose, interface software has been installed at national hydrometeorological institutes of the DRB to provide standardized data services being also a data source for the EFAS. DAREFFORT project is a transnational initiative led by Hungary (VIZITERV Environ Ltd) and received a Letter of Recommendation from the Steering Group of EUSDR PA5. It is financed by the Interreg Danube Transnational Programme. Its preparation was supported by the Hungarian Ministry of Foreign Affairs and Trade.

Meteorological Data

Meteorological observations are an essential part of flood and ice warning and forecasting systems. Generally, within meteorological networks, various data are collected. The most important variables are precipitation, air temperature, air humidity, wind speed, air pressure, solar radiation, sunshine duration, evaporation, soil moisture, snow depth, and snow water equivalent.

It has been identified that actual measurements of evaporation and transpiration are performed only in some countries at a few meteorological stations. Furthermore, potential evaporation is measured only in some places.

Soil moisture measurements are hardly ever taken. Composite weather radar imagery to be used in hydrological models is not available in all countries. In most countries, there are no systematic measurements of the snow water equivalent or its spatial distribution, despite the fact that the floods in DRB are mainly generated in mountainous areas in combination with snowmelt.

Evaluation of Forecasting

The availability and access to meteorological data are different across the countries of the Danube River Basin. Furthermore, not all the data are free of charge in the respective countries. Availability of meteorological data, terms of use for the countries involved in the DAREFFORT project, the number of online meteorological stations operated in the Danube River Basin, and information about the type of data provided on the public website are limited.

Hydrological Data

In the frame of hydrological monitoring, all countries collect data on hydrological parameters, i.e., water level, discharge, and water temperature. Some of them collect information about sediment and ice, while there are practically no systematic measurements of water flow velocity. Bedload transport is hardly ever measured. There are no systematic measurements of channel morphology either, with the exception of navigable waterways along the Danube and its tributaries. Measurements of river stages, and indirect river discharges, are well developed in all hydrological services. However, the number of observation stations has unfortunately decreased over recent decades and we lost valuable information regarding the heterogeneity and dynamics of the phenomena measured. Furthermore, digitalization of historical data is lacking in all services. The availability and access to hydrological data are also different across the DRB countries and not all the data are free of charge in the respective countries. All countries foster the extensive exchange of meteorological and hydrological data and information with domestic and foreign institutions and users. They provide data to international organizations such as the Global Runoff Data Centre (GRDC), the European Flood Awareness System (EFAS), and the Hydrological Information System of the Sava River Basin (SavaHIS). Data are exchanged based on agreements. Procedures for national and international exchange of meteorological and hydrological data exist in all countries.

The countries used their own criteria for selecting the representative hydrological stations on the main rivers and their tributaries in the Danube River Basin proposed for the DanubeHIS. The location of the selected stations is presented in Figure 2.

Ice Data

Ice measurements are conducted along with the Danube River's main flow and its navigable tributaries, based on the recommendation adopted by the Danube Commission. In other water bodies, there are some ice measurements or observations. Because of the frequent problems with ice and historical floods, these services are well organized in Hungary.

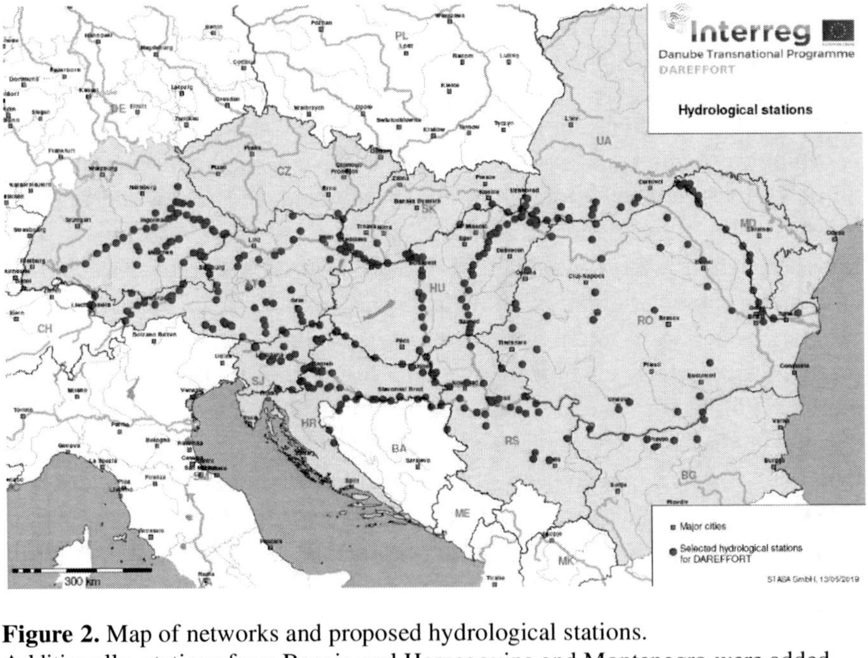

Figure 2. Map of networks and proposed hydrological stations.
Additionally, stations from Bosnia and Herzegovina and Montenegro were added.

Hydrological Forecasting

The type and manner of hydrological forecasting are mostly subject to regional natural conditions. In upper and high-slope river reaches, the discharges mostly depend on the local precipitation type and its intensity. In such regions, monitoring systems with as short as possible data transfer time steps are needed for forecasting as the hydrological conditions are prone to rapid changes. The water levels of the lower river reaches are strongly influenced by the drainage from the upper catchments and the low slope of the river channel. Here, the change of hydrological conditions is slower, and the hydrological conditions are maintained for longer periods. Water levels of low-slope rivers may also be affected by ice.

The national hydrological and the meteorological forecasting services of the Danube River Basin countries mostly operate within the same institution, on a door-to-door principle, which means that hydrological services have access to the meteorological data and predictions free of charge, daily consultations with meteorologists are in place and usually, both services prepare a joint warning product as well. A large diversity among the Danube River countries in terms of hydrological and hydraulic models used and the number of models applied was recognized. Among the hydrological models, the deterministic/conceptual/lumped model type predominates over the deterministic/physically based/distributed model type. The lumped models are often used in a semi-distributed manner by catchment division into sub-catchments and/or elevation zones. The hydrological models used in more than one country (organization) are DHI NAM, HBV/HBV-light, and HEC-HMS. Among the hydraulic/routing models the dynamic wave – hydraulic routing model type predominates over the hydrologic routing model type. Forecasting accuracy assessment is systematically undertaken only by some services, mostly on an occasional basis only. Systematic assessments of forecasts are only performed by the European Commission's Joint Research Centre (JRC) – for its EFAS forecasts, covering EU member states and associate members. The hydrological and hydrodynamic modeling system uncertainty is further weighted with the uncertainty from the meteorological forecasts. Therefore, the hydrological services tend to issue descriptive forecasts rather than quantitative ones originating from the modeling systems. Thus, experienced hydrological forecasters hold a key role in the critical evaluation of the modeling system results as well as within the decision-making processes of the hydrological forecasting service.

Conclusion

Generally, a regular network of meteorological and hydrological gauging stations in the Danube River Basin started to develop in the 19th century. Nowadays almost all countries provide a modern network of hydrological and meteorological stations to ensure real-time data used in forecasting and warning procedures and flood forecasting models.

All countries have an extensive exchange of meteorological and hydrological data and information with domestic and foreign institutions. Hydrological services exchange data and information with neighboring countries for border and cross-border watercourses. The harmonization of flows for border profiles is performed in accordance with pre-defined hydrological criteria and agreements.

Flood forecasting, data collection, and harmonized data sharing are becoming crucially important along the countries of the Danube River Basin. Precipitation, water stage, discharge, water temperature, and ice phenomena are measured isolated in each country.

EU Strategy for the Danube Region Environmental Risks Priority Area initiated a project that is supporting ICPDR in developing the Danube Hydrological Information System. The main aim of the Danube River Basin Enhanced Flood Forecasting Cooperation (DAREFFORT) project is to give a comprehensive overview of the complex meteorological and hydrological measurements and data collection, which has a long history in all countries (more than a hundred years). The main aim of the activity is to enhance access to the recorded hydrological and ice data and to provide harmonized distribution for all the countries in the Danube catchment. For this purpose, for all partners interface software will be installed to provide standardized data services also as the data source for EFAS. DAREFFORT project is a horizontal initiative to implement a flood risk mitigation measure in a joint and sustainable way on the catchment level. Based on the analysis made in the scope of DAREFFORT, recommendations for the improvement of flood and ice forecasting systems were prepared. Some recommendations apply to both meteorological and hydrological data, including ice. Accordingly, the Danube river countries should consider: the need to improve the measuring network in terms of the stations' density and the inclusion of measurements of additional variables; the need for developing meteorological and hydrological products also at the regional level; provision of free access to some of the meteorological and hydrological data related to the national forecasting services; a collection of historical data records, digitalization, and storage in a database; provision of strong arguments for obtaining financial, technical, and human resources to operate meteorological and hydrological services; a common updating interval at least on a daily basis for the parameters. All countries should provide the following hydrological and meteorological parameters for exchange: water level, discharge, water, and air temperature, and precipitation.

References

Centre for Research on the Epidemiology of Disasters CRED. 2019. *Natural Disasters.* Brussels 2020, Vol. 3; p. 5-6.

ICPDR. 2019. Climate change adaptation strategy. *International Commission for the Protection of the Danube River*, Vienna, p. 28-39.

Lawrence J., Blackett P., Cradock-Henry N. A. (2020). Cascading climate change impacts and implications, *Climate Risk Management*, Vol. 29, 100234, https://doi.org/10.1016/j.crm.2020.100234.

Setola R., Theocharidou M. (2017). Modelling dependencies between critical infrastructures. In: *Managing the Complexity of Critical Infrastructures. A Modelling and Simulation Approach*, Setola R., Rosato V., Kuriakides E., Rome E. (Eds.), Springer International Publishing AG, Cham, p. 19-42, https://doi.org/10.10 07/978-3-319-51043-9.

Van Aalst M. (2006). The impacts of climate change on the risk of natural disasters, *Disasters*, 30(1):5-18, https://doi.org/10.1111/j.1467-9523.2006.00303.x.

Zwęgliński T., Smolarkiewicz M., Gromek P. (2020). Efekt kaskadowy współczesnym wyzwaniem zarządzania kryzysowego [*Cascading effect on the contemporary challenge of crisis management*], *Wydawnictwo Szkoły Głównej Służby Pożarniczej, Warszawa*, p. 13-38.

Chapter 6

A Framework for Climate Change Impact Assessment and Mitigation of Hydrological Parameters of the Watershed through Trend Analysis and Hydrological Modeling

Mohdzuned Shaikh[1*],
Pradeep Lodha[2]
and Saeid Eslamian[3]
[1]Department of Civil Engineering
L. D. College of Engineering, Gujarat, India
[2]Department of Civil Engineering
Vishwakarma Government Engineering College, Gujarat, India
[3]Department of Water Engineering, College of Agriculture,
Isfahan University of Technology, Isfahan, Iran

Abstract

Climate change is an inevitable phenomenon that has led the earth to evolve from an ice age to the present era. Due to the temperature rise, the rate of Evapotranspiration (ET) is increasing which leads to a higher rate (chance) of extreme events (floods). This raises the need to analyze the watersheds which show considerable vulnerability to climate change. This study analyses the effect of environmental (Climate) change on Hathmati River which rises from Gujarat Malwa Hills having a total catchment of 1574 sq. km and average annual rainfall of 864mm. The coefficient of variation of annual rainfall over the basin is rather high and ranges between 42% to 65%. The analysis is classified into two parts 1)

---

* Corresponding Author's Email: zunedshaikh@ldce.ac.in.

In: Handbook of Uncertainty in Eurasian Forecasting (HEF)
Editors: Saeid Eslamian, Mir Bintul Huda and Faezeh Eslamian
ISBN: 978-1-68507-966-6

Trend Analysis and 2) Hydrological Modeling. The trend analysis is performed by using Sen's Slope method where significance values for monthly and seasonal data are determined. From trend analysis, it is observed that there exists a considerable positive trend for temperature in accordance with climate change impact. Sen's significance (Z) value ranges from -0.03 to +1.72. The existing trend for rainfall, evapo-transpiration & surface runoff is found to be positive with an average significance value of 2.88, 2.07 & 1.59 respectively. The tool used for hydrological modeling is ArcSWAT (semi-distributed hydrological model) which takes in Geo-Spatial data (and meteorological data) to estimate various parameters such as ET, Runoff, etc. The model run has been carried out for 35 years (1979 to 2014) where model outputs are compared with the observed values of runoff. Sensitivity analysis (Calibration) is performed to calibrate the model through the SUFI-2 (using SWAT CUP) algorithm for the monthly values of runoff giving the average efficiency (NSE) of the model as 0.90. The model is successfully validated for five years giving NSE as 0.89. The Calibrated & validated model shows that the average values of evapotranspiration & surface runoff in mm against 882mm of rainfall are 303mm & 285mm respectively. The analysis has been performed to evaluate the impact of climate change on Evapotranspiration & Surface Runoff through considerable variations in climatic conditions using IPCC scenarios (CMIP-5 projections) and identifying the vulnerable spots that may require attention because of the climate change. The framework is proposed to assess and mitigate the impact of climate change based on the present study.

Keywords: climate change, trend analysis, hydrological modeling, framework, mitigation

List of Abbreviations

Avg.	Average
GIS	Geographical Information System
ha	Hectare
hr	Hour
IRS	Indian Remote Sensing Satellites
km	Kilometer
LISS II	Linear Imaging Self-scanning Sensor II
m	Meter
max	Maximum

min	Minimum
mm	Millimeter
RS	Remote Sensing
SOI	Survey of India
SWAT	Soil and Water Assessment Tool
ArcSWAT	ArcGIS interface Soil and Water Assessment Tool
AMC	Antecedent Moisture condition
HRU	Hydrologic Response Unit
EL	Elevation (m)
LAI	Leaf Area Index
ET	Evapotranspiration
SURQ	Surface Runoff
PET	Potential Evapotranspiration
PCP	Precipitation
Sc	Scenario
TMP	Temperature
CMIP5	Coupled Model Intercomparison Project-5
RCP 4.5	Representative Concentration Pathway

Introduction

Climatic extremes, including heavy precipitation events and extremely hot days, have substantially increased in the past few decades (Alexander 2006; Vose 2005). Changing climate is a continuously occurring phenomenon observed globally (Chandole 2019). It is observed through the change in the statistics of the hydro-meteorological parameters (Chandole 2019). Changing rainfall has a significant effect on water resources, agricultural output, and hence the economy (Oza M. 2014). The effects of hydrological processes and water resources caused by climate change have also received much attention (Gleick, 1986; Burn, 1994). The Assessment of current water availability and its trend in a river basin is of paramount importance to allocate limited water supply for different sectors and to develop strategies for the sustainable use of water in the future (Abeysingha 2015). Further, estimation of water resources of a river basin/sub-river basin in a spatially and temporally explicit way is the key to formulating appropriate policies for equitable and efficient use of limited water resources (Abeysingha 2015). The IPCC predicted that a temperature rise of 2.7-4.3°C over India is expected by the 2080s (IPCC 2007). The precipitation over the Indian sub-continent is expected to rise by

6-8% and sea-level rise by 88 cm by the 21st century (IPCC 2007). Under the A2 IPCC scenario, the mean temperature is predicted to rise by 3-5°C by the end of this century. At a small temporal scale (e.g., daily), hydrological systems are expected to experience not only changes in the average availability of water but also changes in extreme events (Grillakis 2016; Jiang 2007; Modrick 2015; Piras 2016). About a 20% rise in Indian rainfall is predicted by the end of this century. (IPCC 2014). Several studies were partially concentrated on analyzing the impact of climate change and variability on different components of the hydrological cycle (Sonali P. 2012). In the most recent studies, it is observed that significant warming in the second half of the 20th century resulted in a drastic change in the hydrology of an agricultural-based country like India (Sonali P. 2012). Observations over India show that the mean annual surface air temperature has increased by 0.4-0.6°C in the last century (Arora et al. 2005). According to IMD, the all-India annual-mean surface air temperature was the highest in 2016 (IMD 2018). The past thirty years have faced the most increase in temperature during pre-monsoon (March to May) while heavy rainfall event has increased by 50% in central parts of India (Goswami 2006). As the Indian monsoon rainfall simulation is found to be extremely sensitive to initial conditions, some suitable integration procedures have to be devised for seasonal forecasting (World Climate Research Programme 1992). The impact of climate change is projected to have different effects within and between countries (Arora 2005). Information about such change is required at global, regional, and basin scales for a variety of purposes (Arora 2005). There is unmistakable proof of climate change in India in the 20th century that would accelerate in the 21st century (Goswami 2006).

From the literature review of relevant papers, it can be said that more heavy extreme events, sea-level rise, and severe heat waves are expected in the near future. Hence climate change adds up to one more measurement of the issues revealed in today's life.

Case Study and Data

The Hathmati River watershed (23°30′49″N 72°49′29″E) has been selected for the study and the location map is shown in Figure 1. The Hathmati is one of the major tributaries (left) of the Sabarmati River (Western India). The spatial variation in the rainfall is highest for the Hathmati basin among all sub-basins of the Sabarmati basin (Dave H.K 2012). It lies in Bhiloda (Sabarkantha district) and rises from Gujarat Malwa Hills. After traveling a course at 98km it meets Sabarmati near the village Ged. The total catchment area is 1574 sq. km. (157400 ha). The rain gauge station is installed at Bhiloda where the average annual rainfall is recorded as 864 mm. The coefficient of variation of annual rainfall over the basin is rather high and ranges between 42-65% and the average maximum and minimum temperatures reach about 38°C & 16°C (Dave 2012). The data for the study includes various spatial data like DEM, LULC, Soil Map generated from CARTOSAT-I, SRTM, BHUVAN, and IRS~ID LISS III satellite data and prepared in ArcGIS on a 1:12500 scale with the resolution of 30m and several collateral data like weather files (from 1966 to 2015) from rain gauge stations & climate stations have been collected. The spatial maps are shown in Figure 2. The decadal temporal variation from the year 1985 to 2015 in land use has been taken into account and the maps are shown in Figure 3.

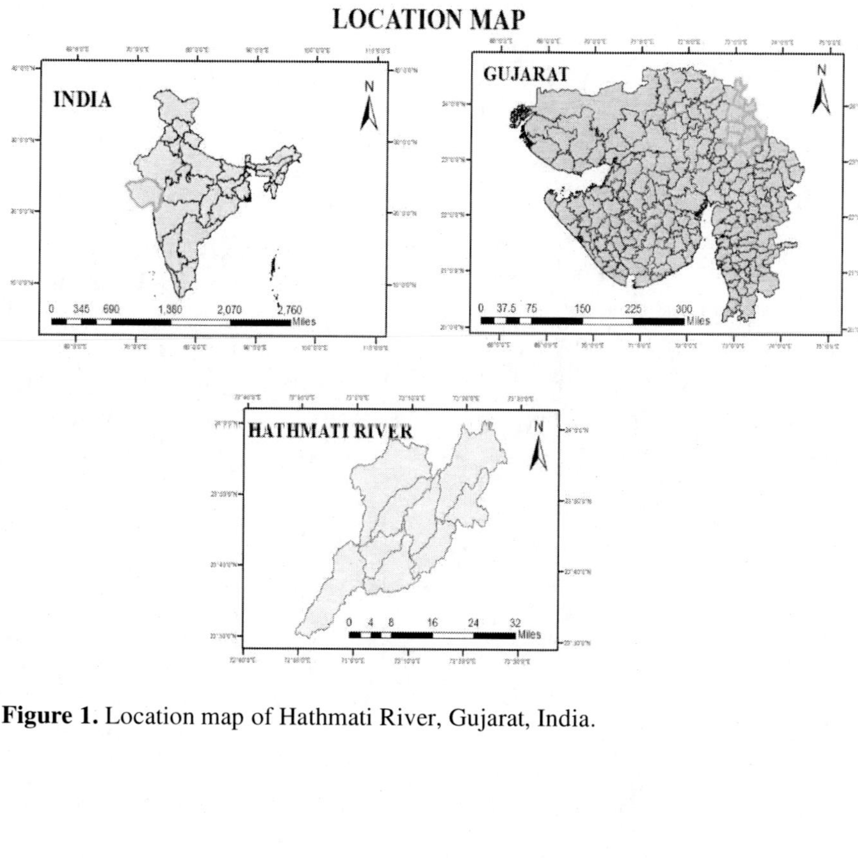

Figure 1. Location map of Hathmati River, Gujarat, India.

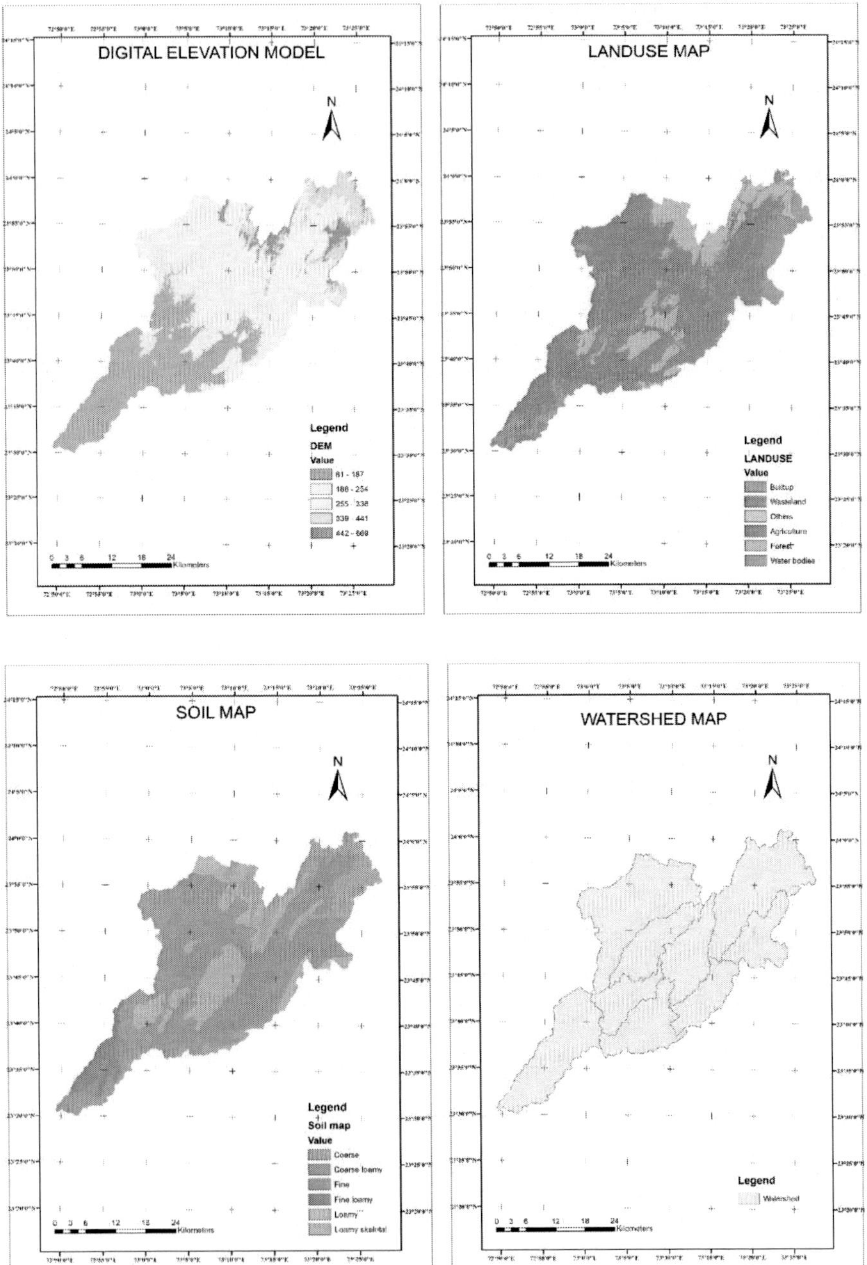

Figure 2. Location map of Hathmati River, Gujarat, India. (Continued).

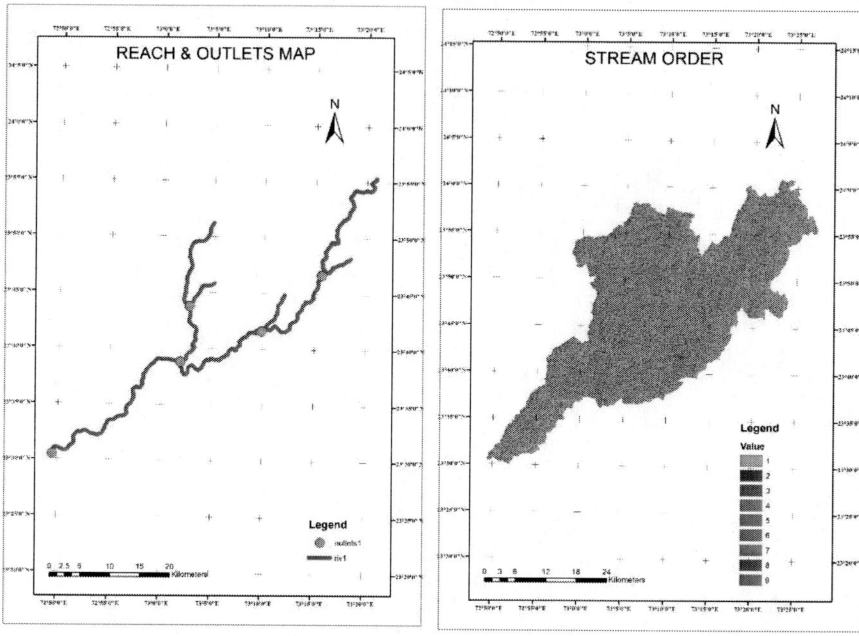

Figure 2. Spatial data of Hathmati River, Gujarat, India.

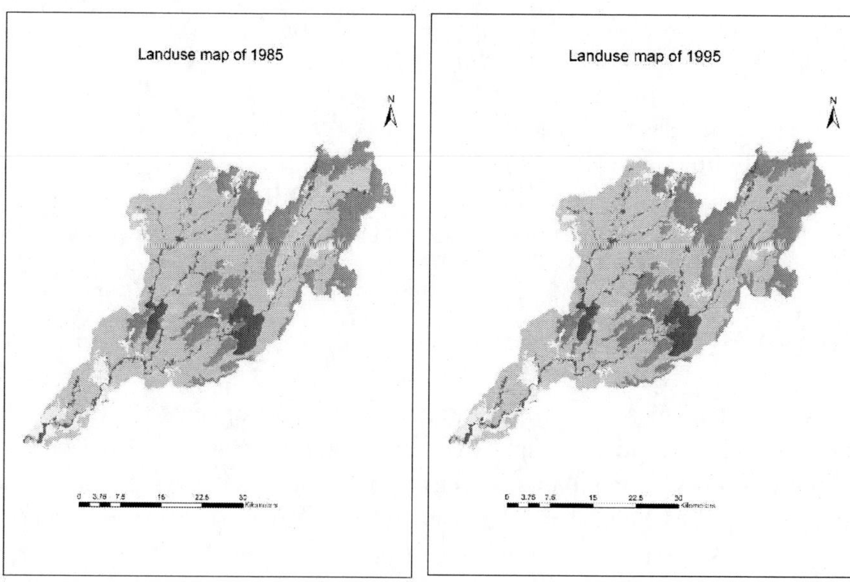

Figure 3. Spatial data of Hathmati River, Gujarat, India. (Continued).

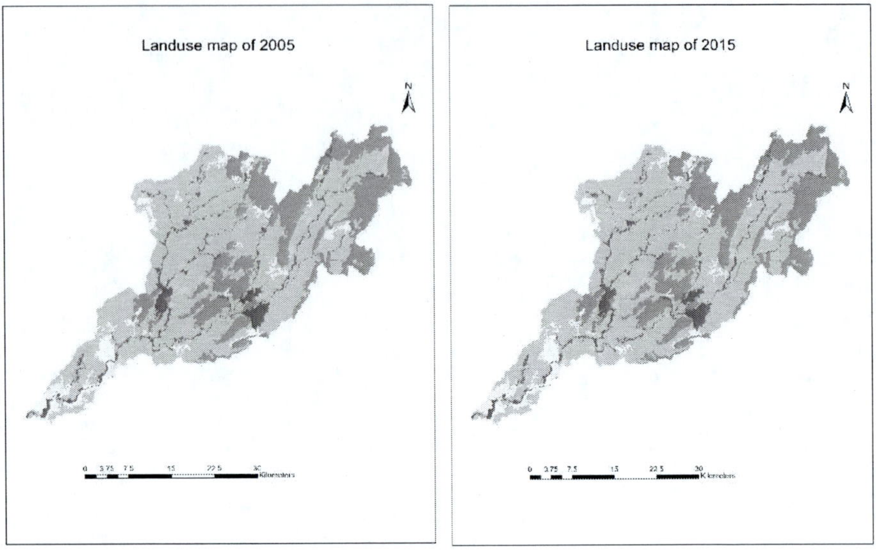

Figure 3. Temporal variation of landcover.

Analysis

The analysis has been done in four major parts:

1. Trend (Statistical) Analysis.
2. Hydrological Modeling.
3. Simulation using Forecasting scenarios of IPCC.
4. A framework to assess and mitigate the impact of climate change.

Trend Analysis

A persistent increase in carbon dioxide in the atmosphere, which may be caused by industrialization and deforestation, has been recorded since the 1950s (Yu 2002). This increased trend may lead to changes in global and regional climate features, such as average temperature and precipitation (Yu 2002). The Magnitude and trend of warming of India during the last century over the Indian continent is matching with the global condition (Pant and Rupa Kumar 1997). The understanding of past climate-ecosystem coupling is thus vital to predicting future ecosystem responses to changing natural variables

(Basu 2018). The analysis commences with performing trend analysis over more than 50 years for temperature & rainfall using Non- Parametric test (Sen's Slope method). To estimate the true slope of an existing trend (as a change per year) Sen's nonparametric method is used. Sen's method can be used in cases where the trend can be assumed to be linear.

This means that f(t) is equal to:

$$f(t) = Qt + B \tag{1}$$

where Q is the slope and B is a constant. To get the slope estimate Q in equation (1) we first calculate the slopes of all data value pairs as:

$$Q_i=(x_j- x_k)/(j-k) \text{ where } j>k. \tag{2}$$

If there are n values xj in the time series, we get as many as N = n(n-1)/2 slope estimates Qi. Sen's estimator of the slope is the median of these N values of Qi.

The N values of Qi are ranked from the smallest to the largest and Sen's estimator is:

$$\begin{gathered} Q=Q_{[(N+1)/2]}, \text{ (if N is Odd)} \\ Q=0.5(Q_{[N/2]}+ Q_{[(N+2)/2]}, \text{ (if N is Even)} \end{gathered} \tag{3}$$

A 100(1-α)% two-sided confidence interval about the slope estimate is obtained by the nonparametric technique based on the normal distribution. The method is valid for n as small as 10 unless there are many ties. The method computes the confidence interval at two different confidence levels; $\alpha= 0.01$ and $\alpha = 0.05$, resulting in two different confidence intervals.

At first, we compute Cα as:

$$C\alpha=Z_{(1- \alpha)}[VAR(S)]^{0.5} \tag{4}$$

where VAR(S) has been defined as:

$$VAR[S] = \frac{1}{18}[n(n-1)(2n+5) - \sum_{p=1}^{q} tp(tp-1)(2tp+5)] \tag{5}$$

Next M1 = (N - Cα)/2 and M2 = (N + Cα)/2 are computed. The lower and upper limits of the confidence interval, Q_{min}, and Q_{max} are the M1th largest and the $(M2 + 1)^{th}$ largest of the N ordered slope estimates Qi. If M1 is not a whole number the lower limit is interpolated. Correspondingly, if M2 is not a whole number the upper limit is interpolated. To obtain an estimate of B in equation (1) the n values of differences $x_i - Q_{ti}$ is calculated. The median of these values gives an estimate of B (Sirois 2013). The estimates for the constant B of lines of the 99% and 95% confidence intervals are calculated by a similar procedure. The present paper shows the trend generated for a) Linear regression b) Sen's estimate c) at max 95% confidence level d) at min 95% confidence level.

Table 1. Summary of trend analysis

Month	Significance Value Z for Mean Temperature	Significance Value Z for Maximum Temperature	Significance Value Z for Minimum Temperature
Jan	1.32	1.56	1.61
Feb	-0.39	-1.3	-1.29
Mar	-0.19	-0.77	-0.74
April	0.66	1.3	1.24
May	1.06	1.46	1.36
June	0.86	0.96	0.94
July	-0.21	0.35	0.39
Aug	0.26	1.47	1.36
Sep	1.73	3.09	3.19
Oct	0.03	2.93	3.04
Nov	0.58	2.33	2.11
Dec	1.45	2.46	2.54
Average	1.33	2.93	3.06

The research work is initiated by collecting the observed weather data over more than 50 years for the said study area and statistical analysis is performed. This analysis is used to determine the trend of rainfall and temperature over the temporal variation and the results are used to generate scenarios in Hydrological Modeling (ArcSWAT). From the graphs, it is clearly understood that trend for temperature and rainfall is positive. Hence there is clear evidence of climate change and it will accelerate to cause severe effects on the hydrological parameters.

Detail trend analysis is done by Non-parametric tests using Sen's Slope method which on temperature & rainfall over 50 years (1966-2015). This approach is not valid when the data elements don't fit a straight line; it is also sensitive to outliers (Wang 2005). The detailed trend analysis clearly shows the evidence of the impact of climate change on the Hathmati river basin. Except for Feb, March, and July, each month shows a considerable positive trend toward the temperature rise ranging from -1.29 to +3.09. The trend for annual mean rainfall shows a considerable positive trend (Z=1.33) as well. This trend is utilized to generate scenarios for different climatic condition and their impact.

Hydrological Modeling

Rainfall-runoff modeling can be carried out within a purely analytical framework based on observations of the inputs to and outputs from a catchment area (Beven 2012). The catchment is treated as a black box, without any reference to the internal processes that control the transformation of rainfall to runoff (Beven 2012). While the issues of deterministic predictability of the monsoon rainfall on a seasonal scale are yet to be resolved, optimizing the accuracy of the forecast based on empirical models by a proper selection of the predictands, predictors, and methodology has immediate practical applications (Rupa 2002). Model construction, calibration, and data processing have received a great deal of attention, while model validation, error propagation, and analyses of uncertainty, risk, and reliability have not been treated as thoroughly (Singh 2002). SWAT is the acronym for the Soil and Water Assessment Tool, a river basin, or watershed, scale model developed by Dr. Jeff Arnold for the USDA Agricultural Research Service (ARS). SWAT was developed to predict the impact of land management practices on water, sediment, and agricultural chemical yields in large complex watersheds with varying soils, land use, and management conditions over long periods (Neitsch et al. 2001). Hydrologic analysis of large watersheds draining thousands of square kilometers is a tedious process (Jayakrishnan 2005). Regional and national-scale water resources management strategies and decision-making for sustainable domestic, agricultural, and industrial water supply, as well as protection of the environment from the negative impacts of developmental activities, depend on such large-scale analyses (Jayakrishnan R 2005). Computer-based environmental simulation models like SWAT are valuable tools for such studies (Jayakrishnan 2005). Since hydrologic systems are heterogeneous, with substantial spatial variability in model inputs such as soil and land use, the development of input files for such models for large watersheds is a time-consuming task (Jayakrishnan 2005). GIS plays a major role in developing model inputs from digital geospatial databases through model–GIS interfaces (Jayakrishnan 2005). The SWAT model and its GIS interfaces aid the water resources professional in basin-scale studies of water availability and water quality and help reduce the time and cost necessary to conduct such studies several-fold compared to other distributed parameter models (Jayakrishnan 2005). The Global and regional climate models (GCMs/RCMs) do not represent valley/ridge scale interactions well (Shafiq 2019). The SWAT model has good potential for application in hydrologic/water quality studies in countries around the world and as a tool to develop time and cost-efficient analyses for watershed/water resources management and decision-making (Jayakrishnan 2005). Rather than incorporating regression equations to describe the relationship between input and output variables, SWAT requires specific information about weather, soil properties, topography, vegetation, and land management practices occurring in the watershed (Neitsch 2001).

The SWAT is not an uncommon tool and hence the detailed procedure to run the model is not explained in the paper. Broadly the model run is classified into four stages 1) Watershed delineation 2) HRU formation 3) Weather files input 4) Model run (Simulation). The necessary data required to set up the model includes spatial maps and temporal weather data which are already discussed earlier. The heart of any modeling work is to calibrate the model which is discussed next.

Model Calibration (Sensitivity Analysis)

For hydrological prediction, such as discharge, careful model calibration is required for an efficient result (Narsimlu 2015). For a good modeling practice, it is required to report the uncertainties in the model prediction along with the results (Narsimlu 2015). The model calibration is carried out from 2000 to 2010 on monthly basis. The study aims to calibrate the model to resemble the existing watershed in all the hydrological ways. The model is calibrated by the "Sequential Uncertainty Fitting" (SUFI-2) algorithm available as an interface with SWAT CUP where surface runoff over the reach is taken as a variable and four sensitive parameters are chosen. The SWAT-CUP is specially developed by Abbaspour et al. (2007) to interface with the SWAT model. Several objective functions have been used for estimating model performance, which includes the Root Mean Square Error (RMSE), the absolute difference, and the logarithm of differences, the R^2, the Chi-square, and the Nash-Sutcliffe Efficiency (NS). Several objective functions have been utilized to reduce the non-uniqueness problem in model characterization (Duan et al. 2006). The study aims to calibrate the model to resemble the existing watershed in all the hydrological ways. The model is calibrated by the "Sequential Uncertainty Fitting" (SUFI-2) algorithm available as an interface with SWAT CUP where surface runoff over the reach is taken as a variable and six sensitive parameters as shown in Table 2 are chosen.

Table 2. Absolute values of sensitive parameters after calibration

Sensitivity Rank	Swat Parameter Name	Swat Parameter Description	Unit	Range	Absolute Value
1	CN2	Curve number	-	30-98	75
2	ALPHA_BF	Baseflow recession constant in days.	day^{-1}	0-1	0.7
3	GW_DELAY	Delay time for aquifer recharge in days	days	0-500	240
4	Soil_Z	Soil depth	mm	0-3500	300
5	GWQMN	Threshold level for baseflow	mm	0-5000	1.0
6	SURLAG	Surface runoff lag time	-	0.05-24	0.1

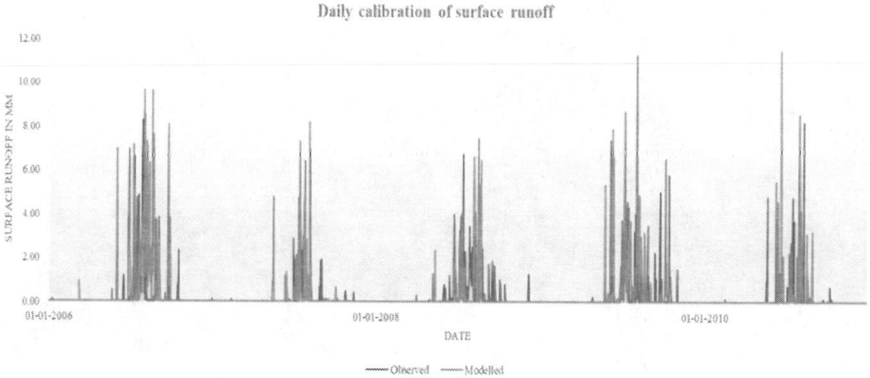

Figure 4. Daily calibration from 2006 to 2010.

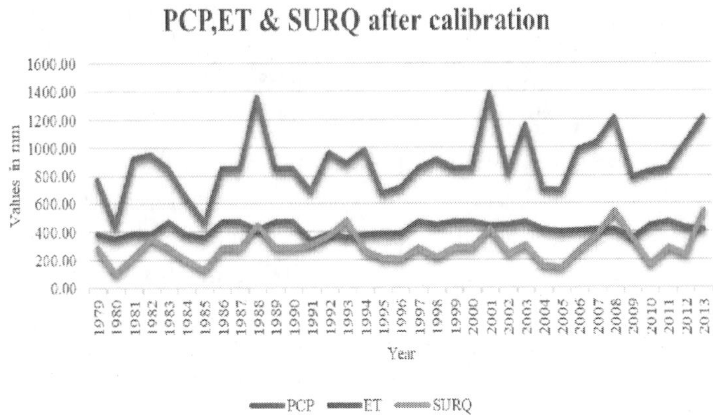

Figure 5. Average monthly values of precipitation, ET & surface-runoff after calibration.

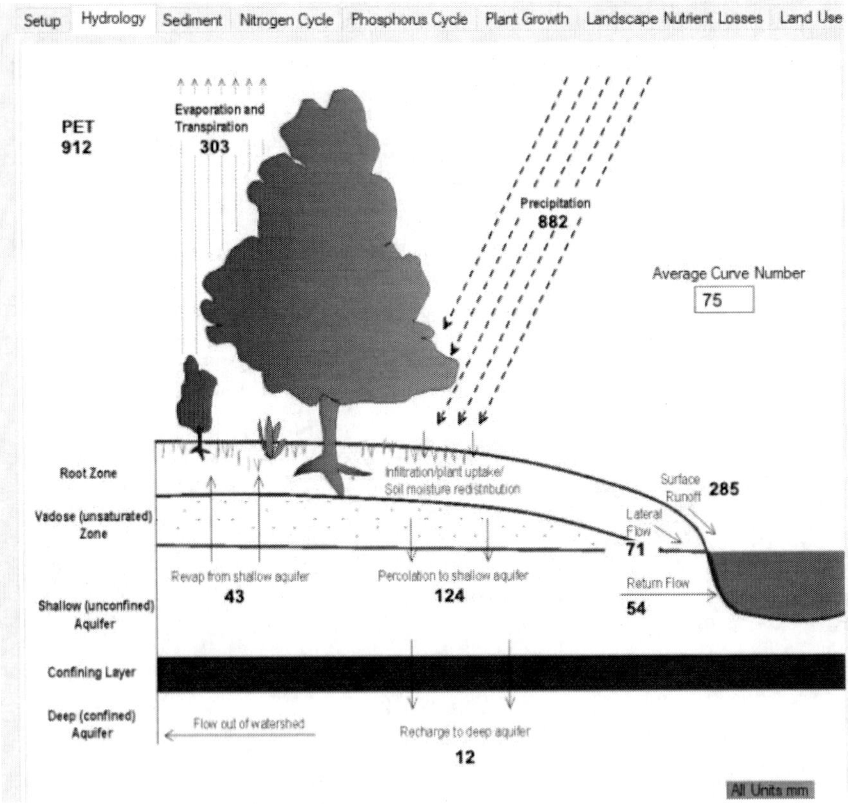

Figure 6. Hydrology of the Hathmati watershed from SWAT output.

After successfully completing the model run, the results are summarised as:

- Semi-distributed model SWAT is used to find ET
- The model is calibrated/validated with observed values
- Simulation over 36 years is generated to find variation in ET
- SURQ contributes about 28-30% while ET losses are about 35%
- NSE after calibration & validation is obtained at around 0.89
- The trend for ET and rainfall is found significantly positive
- The model shows the average value of rainfall, evapotranspiration, surface runoff, lateral flow, baseflow, and infiltration in mm as shown in Table 5.2 as 882, 303, 285, 71, 124, and 193 respectively

The outputs from the SWAT model are shown in Figure 5 to Figure 7:

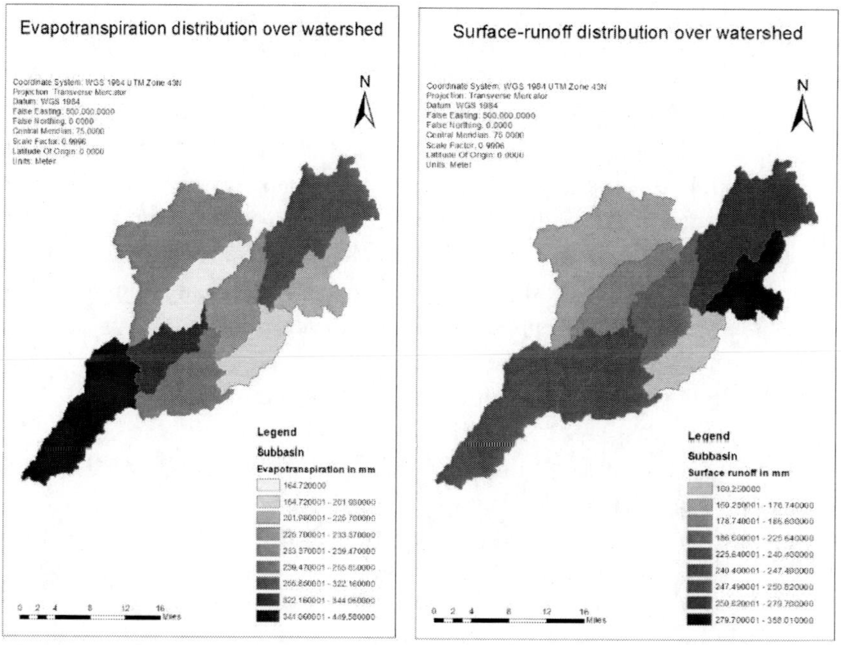

Figure 7. Maps of outputs after calibration.

Simulation Using Forecasting Scenarios of IPCC (CMIP-5 Projections)

Watershed models are often used to simulate the impact of future climate conditions on hydrologic processes. However, Teutschbein and Seibert (2012) state that simulations of temperature and precipitation often show significant biases due to systematic model errors or discretization and spatial averaging within grid cells, which hampers the use of simulated climate data as direct input data for hydrological models. Bias correction procedures are used to minimize the discrepancy between observed and simulated climate variables on a daily time step so that hydrological simulations driven by corrected simulated climate data match simulations using observed climate data reasonably well. In this study, the coupled model interpretation program (CMIP-5) has been used for generating forecasting scenarios. The tool used for this task is CMhyd (Climate Model data for hydrologic modeling) which is available in the public domain. CMhyd is a tool that can be used to extract and bias-correct (downscale) data obtained from global and regional climate models. The selection of the model is done as per the recommendations of (Jena p. et al. 2015). Their study shows that four models, CCSM4, CESM1-CAM5, GFDL-CM3, and GFDL-ESM2G, best capture the pattern in Indian summer monsoon rainfall over the historical period (1871–2005). For assessing the present and future climate change scenarios, daily maximum temperatures from the CMIP5 climate model historical simulations and projections based on Representative Concentration Pathways (RCP4.5) scenario (Taylor et al. 2012) were also used.

The whole process of this simulation can be summarised as:

1. Find the existing anomalies between historical observed data and modeled data (CMIP5).
2. Do the bias correction using the CMhyd tool to minimize the anomalies for all the selected models.
3. Select the model showing the best fit or minimum anomalies.
4. Do the future projections based on this model using Representative Concentration Pathways (RCP4.5) scenario
5. Incorporate the data of temperature and rainfall in the model and extract the outputs for runoff and evapotranspiration.

Table 3. Details of CMIP5 models considered for the study

Sr. No.	Model (Lat^{0}x Long^0)	Modeling Centre (Group)	RMSE
1	CCSM4 (1.25 × 0.9424)	National Center for Atmospheric Research, U.S.A	0.74
2	CESM1-CAM5 (1.25 × 0.9424)	Community Earth System Model Contributors, NCAR USA	0.59
3	GFDL-CM3 (2.5 × 2)	NOAA Geophysical Fluid Dynamics Laboratory	-
4	GFDL-ESM2G (2.5 ×1.5169)	NOAA Geophysical Fluid Dynamics Laboratory	-

From the extensive literature review, the selection of models is done and the details of the models used are shown in Table 3. All the models are compared to the historical observed data and the bias correction is done using the CMhyd tool which is available in the public domain. The results of the comparison of all the four models with the observed historical data after the bias correction show that CCSM4 is the most efficient model. The comparison of this model with the observed data is shown in Figure 8 and the future projections obtained from this model are used as input data for the model. The projected values of ET and surface runoff are shown in Figure 9 and Figure 10.

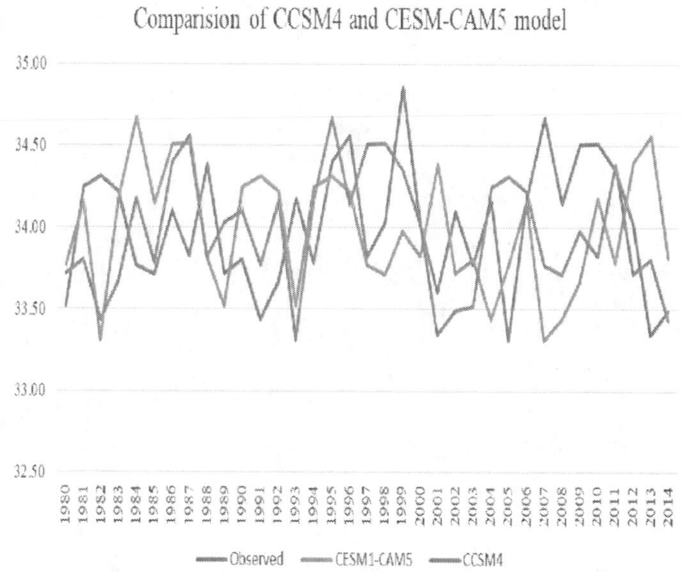

Figure 8. Comparison of CCSM4 and CESM1-CAM5 model for efficiency.

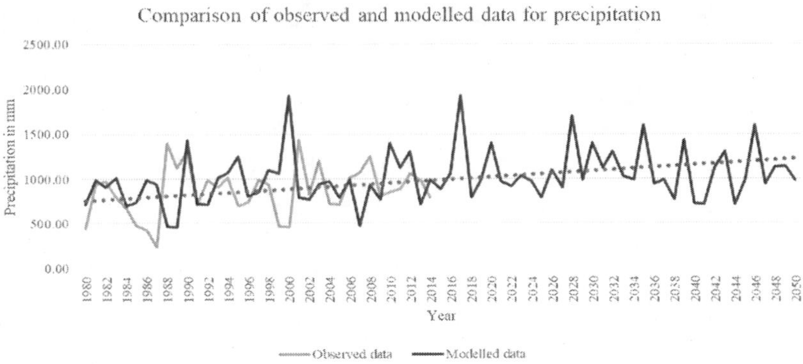

Figure 9. Comparison of CCSM4 and CESM1-CAM5 model for efficiency (Continued).

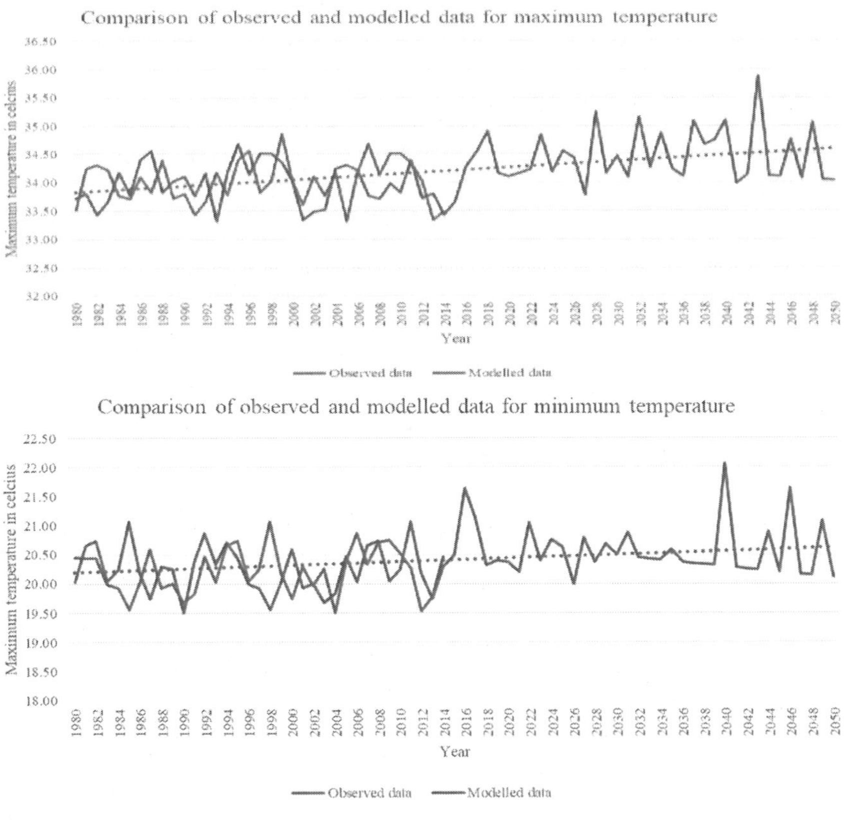

Figure 9. Comparison of CCSM4 model with observed data along with the RCP-4.5 projections.

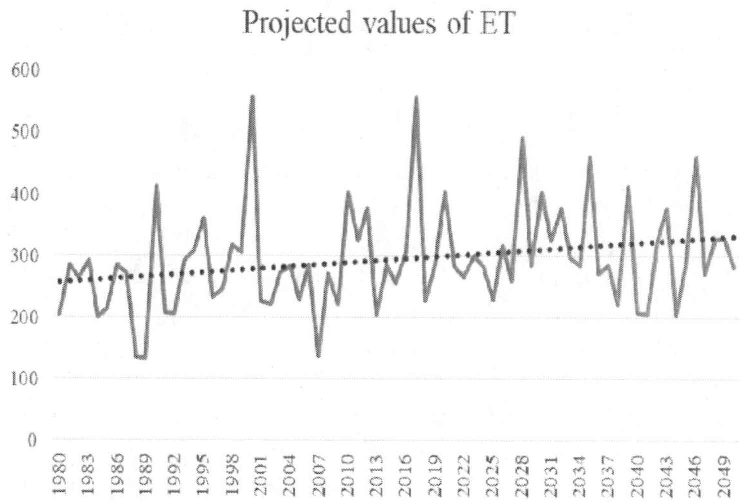

Figure 10. Comparison of CCSM4 model with observed data along with the RCP-4.5 projections. (Continued).

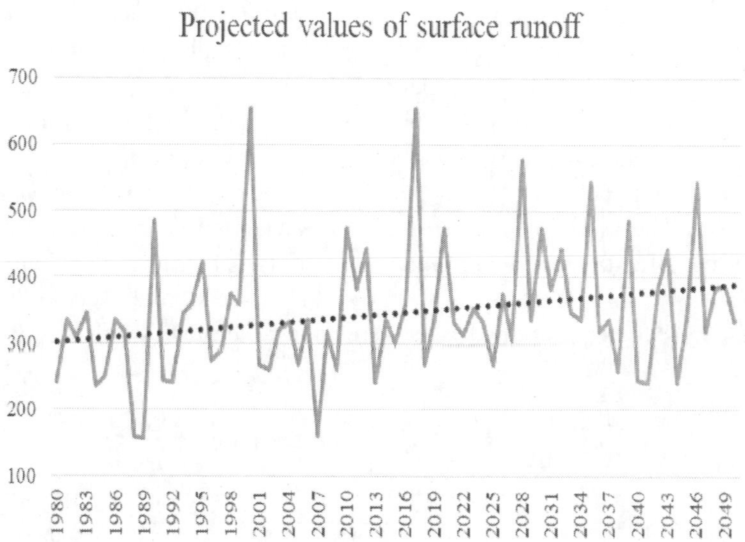

Figure 10. Projected values of ET and surface runoff.

A Framework to Mitigate the Impact of Climate Change

The impacts of climate change are of particular concern to the coastal region of tropical countries like India, which are exposed to cyclones, floods, tsunamis, seawater intrusion, etc. (Krishnan 2018). Climate-change adaptation presupposes a comprehensive assessment of vulnerability status (Krishnan 2018). In developing countries where runoff records are seldom available in medium and minor projects, the water balance coupled with a spatially distributed modeling approach can be used for generating future scenarios of reservoir operation under changes in climate, land use, and other developmental activities (Jaiswal 2020). The policymakers need to be aware of the different characterizations of extremes, and that characterization of extremes requires both interpretability and statistical rigor (Ghosh 2012). The study went through various analyses that dealt with trends of weather data and simulation of existing data to assess the possible outcomes in the near future. The results of the scenarios clearly show the evidence of climate change's impact on evapotranspiration and runoff making the watershed more vulnerable to global warming. The scenarios of IPCC AR-4 have been incorporated for the simulation and each of the simulation results witnesses the existence of a considerable increase in the magnitude of extreme events. This module of the study presents (proposes) the measures to mitigate the impact of climate change by enhancing the green cover of the watershed. The entire work has been represented in the form of a framework as shown in Figure 11. The mitigations are the precautionary measures to overcome the occurrence of extreme events. In the study, only one criterion i.e., enhancing the landcover in the favor of hydrology has been proposed. While the framework can be developed by several other measures like changing crop patterns etc.

Table 3. Land-cover distribution

Features	1985	1995	2005	2015	New
Agriculture	57	59	60	61	62
Forest	25	24	24	26	33
Water	6	2	2	3	2
Built-up	9	9	7	1.6	3
Wasteland	2	2	2	7	
Others	1	4	5	2	

From the existing landcover, the mixed forest seems to dominate (after agriculture) with a score of about 26% while in this module of setting up the framework, it has been enhanced to around 33% evacuating all the wasteland and settlements. So the new landcover that is developed through GIS is shown in Table 3.

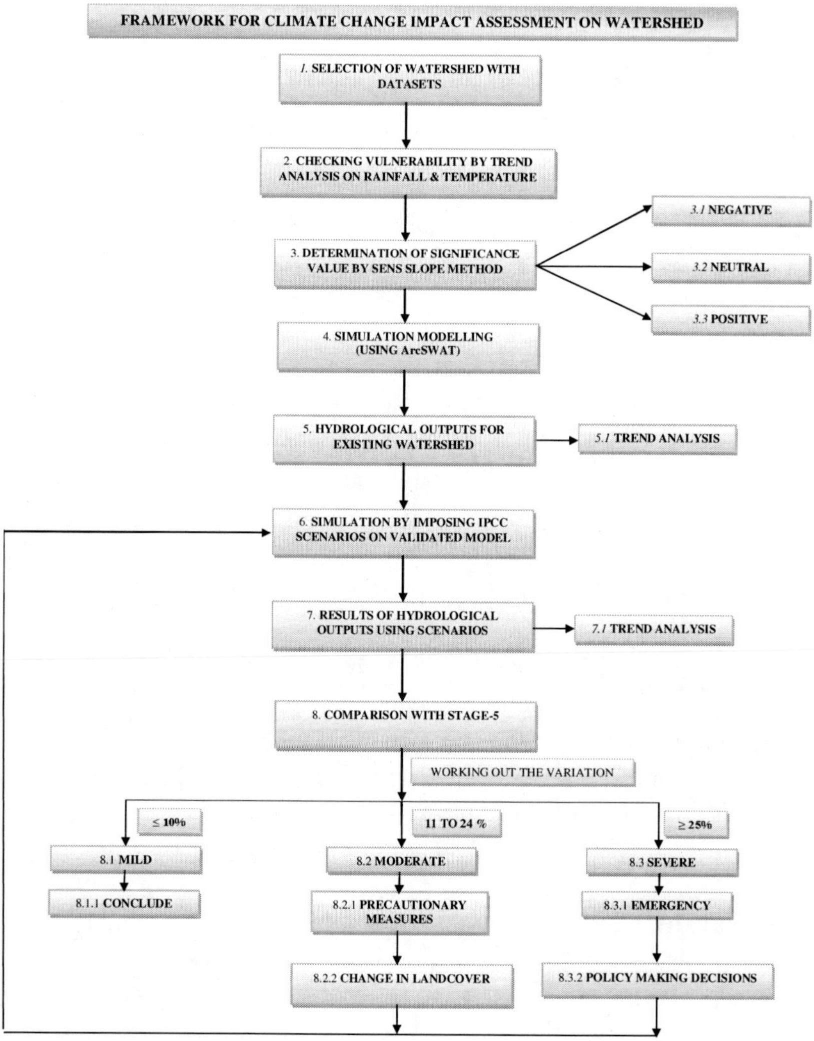

Figure 11. Framework for climate change impact assessment.

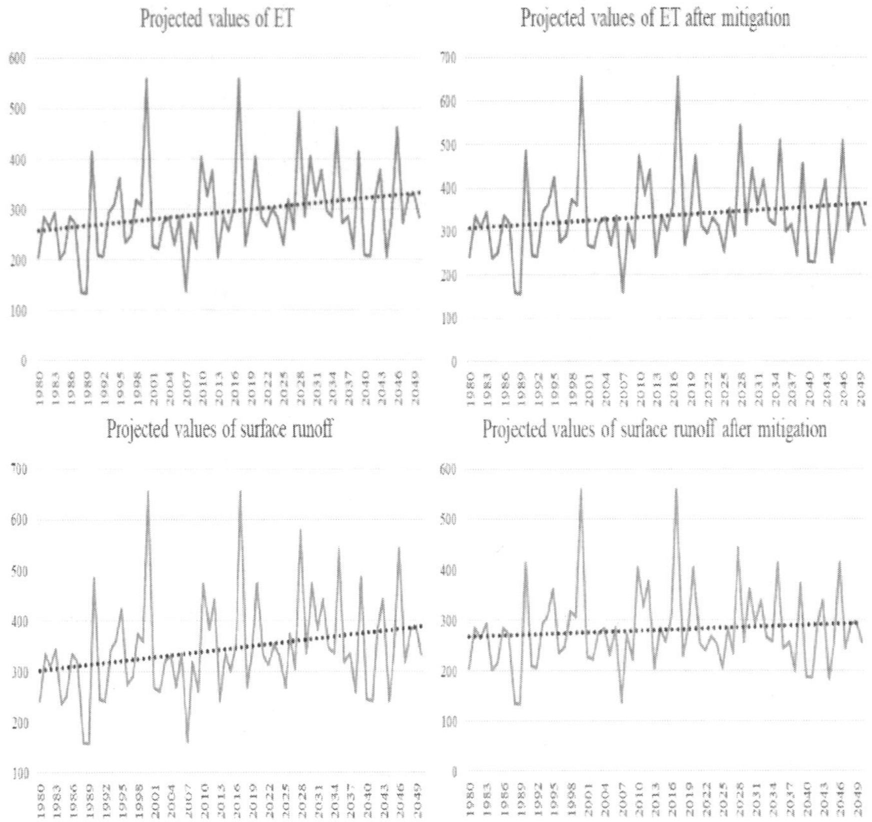

Figure 12. Comparison of projections of evapotranspiration and runoff before and after mitigation.

All the input data to be incorporated in the model are left unaltered except the landcover and the simulation is done for both scenarios. The results in statistical as well as a graphical format are shown in Figure 12.

Results and Discussion

The present work estimates the impact of climate change on hydrological parameters (Evapotranspiration and Surface Runoff) through hydrological modeling. The analysis has been performed to evaluate the impact of climate change on ET & Surface Runoff through considerable variations in climatic conditions (IPCC scenarios) and thus identifying the vulnerable spots that may require attention in view of the climate change.

The work is initiated by the trend analysis followed by hydrological modeling that introduces SWAT in association with ArcGIS to form a simulated watershed. SWAT is chosen as the model for the study, as it is free, open-source, actively supported by model developers, well documented, and has been tested around the world. The SWAT interface was written as a plug-in for ArcGIS which is the way most dynamic existing SWAT module. ArcGIS version 10.2.2 interfaces with SWAT 2012 is used for the study and like any other semi-distributed hydrological model it acquires several spatial data in the form of maps and some temporal data like weather files. The model is calibrated by choosing parameters (Variables) affecting the value of surface runoff (Considerably) for the period of 2000 to 2010 (10 years). The parameters affecting the value of surface runoff are calibrated as a part of sensitivity analysis and the results are obtained giving the average NSE as 0.90. The model is validated for the next five years (2011-2014) and monthly values are shown below giving NSE as good as 0.88. Figure 6 gives a summary of Hydrological Modeling after calibration & validation of the model giving an annual average value of PCP, ET, and SURQ in mm as 882, 303 & 285. Surface runoff contributes about 28-30% while ET losses are about 35%.

The calibrated model is used to commence the simulation using forecasting scenarios of IPCC. This objective is served by using the CMIP5 project database and the models used for the watershed are shown in Table 3. From the extensive literature survey on the CMIP5 models, two models namely CCSM4 and CESM1-CAM5 are selected to process the bias correction. As shown in Figure 8 both the models are compared after doing the bias correction in the tool named CMhyd. From the RMSE score of both models, CCSM4 is chosen to run the projections. For projections, the RCP 4.5 project used where the precipitation and temperature are projected up to 2050 as shown in Figure 9. These data are incorporated in SWAT (already calibrated) and the outputs in the form of ET and surface runoff are worked out. These outputs are plotted and respective significance values are obtained after the trend analysis. The numbers (significance values) as shown in Table 4 show the considerable increase in the trend if IPCC scenarios are executed in the model. As presented in Figure 12, a framework has been established to mitigate the impacts in terms of ET and runoff. One of the measures adopted in the study is the enhancement in the landcover (LULC) and simulation is performed. The simulation result shows a considerable change in the magnitude of hydrological parameters and significance values. For ET the projected value goes down by almost 8 to 9% while for SURQ the same value is about 7.5 to 8%.

From the past research and reports (IPCC) relevant to climate change and its impact on the watershed, the study area can be defined that seems to be more vulnerable to the effect of climate change, followed by the relevant data collection and compilation. The dataset would include spatial and temporal data in the form of maps and statistical format. This can be done through government bodies like data centers and space organizations. More specifically the spatial data required for the study would include a digital elevation model (DEM) of the tributaries, land-use/landcover map, soil map, weather stations map, climate station map, etc whereas temporal data includes weather data like rainfall, temperature, humidity, solar radiation, and wind velocity. These data are then used to get a better idea about the vulnerability of the watershed through trend analysis. The temporal data collected/compiled are then analyzed in this part of the framework. This is done by performing trend analysis over the climatic data, more specifically rainfall and temperature data. According to IPCC temperature and rainfall are the two most important climatic parameters to find the impact of climate change on hydrological parameters. The trend analysis can be performed by various methods which are explained in the next stage of the framework. The more popular methods to perform trend analysis are the Mann-Kendall test and Sen's slope method. The literature review has provided much evidence about the research including these methods for statistical analysis of the data. This study uses Sen's slope method to determine the existing trend of the climatic condition of the watershed. The method is a linear regression that formulates the slope of the trend line in the form of a parameter called a Significance value designated as Z. The range of this value helps to decide the trend of the data. Closer to zero is the value (of Z), and lesser is the vulnerability of the watershed in the context of climate change. Accordingly, positive values show a positive trend leading to a higher probability of occurrence of maximum events while negative values evident the occurrence of minimum events soon. The statistical analysis is followed by the modeling part where spatial data in the form of maps and collaterals are incorporated into the hydrological model to determine the value of the hydrological parameters of the watershed. Several hydrological models are available to serve the purpose but it is recommended to go for distributed or semi-distributed models that can vary spatially and/or temporally. In this study, the model used is SWAT (soil and water assessment tool) which is a semi-distributed model allowing the data to vary spatially. Selection of the model is also derived from the review of past research in the relevant field which has proved the model to give substantially positive results. The model works as an interface of GIS wherein ArcGIS is used for the study. The model is first calibrated and validated by comparing the results with the observed value of the outputs followed by the simulation. Detailed results are shown in Table 4.

Table 4. Result of significance values of ET and runoff before and after mitigation

Parameters	Annual Average of Existing 36 Years	Projected Value in the Year 2050 With Existing LULC	Projected Value in the Year 2050 With New LULC	Sen's Slope Z (Existing LULC)	Sen's Slope Z (New LULC)
ET	303	355	340	2.67	2.61
SURQ	285	320	308	1.59	1.41

Conclusion

The study deals with Climate change which happens to be an inevitable process that has apparently led the earth to evolve from an ice age to the present era. The natural elements such as the atmosphere, hydrosphere, geo-sphere, and biosphere along with human factors like land and its use cause local, regional, or even global changes in climate. It is a prevailing crisis with health and safety, economic, security food production, etc. The Hathmati River watershed has been selected for the study which is one of the major tributaries (left) of the Sabarmati River. The spatial variation in the rainfall is highest for the Hathmati basin among all sub-basins of the Sabarmati basin. The river happens to be one of the most important tributaries for western India following climate change.

The study went through various analyses that dealt with trends of weather data and simulation of existing data to assess the possible outcomes in the near future. The results of the scenarios clearly show the evidence of climate change's impact on evapotranspiration and runoff making the watershed more vulnerable to global warming. The scenarios of IPCC (CMIP5) have been incorporated for the simulation and each of the simulation results witnesses the existence of a considerable increase in the magnitude of maximum events.

The simulation of scenarios developed in the previous module shows considerable results in accordance with the climate change impact. There arises an inevitable necessity to mitigate these vulnerable spots in the channel of optimizing the adversities, ignorance of which may lead to the failure of the natural watershed so developed in the watershed. The present module provides the framework that enhances the landcover in favor of the watershed while the other measures i.e., changing crop patterns and alteration in soil data are left under the scope of the study. As far as the objectives of the study are concerned, landcover plays a vital role in the analysis of hydrological parameters. Though the forest provides the major contribution to the landcover, enhancement in the green cover could change the scenarios affirmatively. From the existing landcover, the mixed forest seems to dominate (after agriculture) with a score of about 26% while in this module of setting up the framework, it has been enhanced to around 36% evacuating all the wasteland and settlements. The simulation is performed by keeping the rest of the inputs unaltered.

A framework has been established to mitigate the impacts in terms of ET and runoff. From the past research and reports (IPCC) relevant to climate change and its impact on the watershed, the study area can be defined that seems to be more vulnerable to the effect of climate change, followed by the relevant data collection and compilation. The dataset would include spatial and temporal data in the form of maps and statistical format. This can be done through government bodies like data centers and space organizations. More specifically the spatial data required for the study would include a digital elevation model (DEM) of the tributaries, land-use/landcover map, soil map, weather stations map, climate station map, etc whereas temporal data includes weather data like rainfall, temperature, humidity, solar radiation, and wind velocity. These data are then used to get a better idea about the vulnerability of the watershed through trend analysis. The temporal data collected/compiled are then analyzed in this part of the framework. This is done by performing trend analysis over the climatic data, more specifically rainfall and temperature data. According to IPCC temperature and rainfall are the two most important climatic parameters to find the impact of climate change on hydrological parameters. The trend analysis can be performed by various methods which are explained in the next stage of the framework. The more popular methods to perform trend analysis are the Mann-Kendall test and Sen's slope method. The literature review has provided much evidence about the research including these methods for statistical analysis of the data. This study uses Sen's slope method to determine the existing trend of the climatic condition of the watershed. The method is a linear regression that formulates the slope of the trend line in the form of a parameter called a Significance value designated as Z. The range of this value helps to decide the trend of the data. Closer to zero is the value (of Z), and lesser is the vulnerability of the watershed in the context of climate change. Accordingly, positive values show a positive trend leading to a higher probability of occurrence of maximum events while negative values evident the occurrence of minimum events in the near future. The statistical analysis is followed by the modeling part where spatial data in the form of maps and collaterals are incorporated into the hydrological model to determine the value of the hydrological parameters of the watershed. Several hydrological models are available to serve the purpose but it is recommended to go for distributed or semi-distributed models that can vary spatially and/or temporally. In this study, the model used is SWAT (soil and water assessment tool) which is a semi-distributed model allowing the data to vary spatially. Selection of the model is also derived from the review of past research in the relevant field which has proved the model to give substantially positive results. The model works as an interface of GIS wherein ArcGIS is used for the study. The model is first calibrated and validated by comparing the results with the observed value of the outputs followed by the simulation.

The hydrological outputs like evapotranspiration and runoff are obtained for all the years and trend analysis is performed on these outputs the same way as done in stage 2 and stage 3. The significance value of hydrological parameters helps to decide the consequences of the impact of climate change in the near future. What could be the scenario soon if the existing trend persists, can be proposed from this analysis. Once the model is calibrated and validated, it can be graded as capable to simulate the outputs in different scenarios. More specifically, the model can be used to project the outputs in the form of hydrological parameters under the different climatic conditions in the near future.

This study incorporates the scenarios suggested by IPCC (CMIP5 and RCP4.5 projections) where considerable changes are done in the weather files and projections are made up to the year 2050.

The results from the model in the form of hydrological parameters are summarised and trend analysis is performed on the outputs. This module provides two significant outputs as the projected value of hydrological parameters (say runoff) and the significance value of the trend line generated from the trend analysis. These results help to have a comparative analysis between existing and projected values. This stage helps to decide the condition of the watershed during the projections, whether it would go mild, moderate, or severe as compared to the existing values in the form of hydrological parameters. This is done by setting up a range of variations for the values of hydrological parameters and Sen's slope. In this study, the range of variation for the mild condition is proposed to be less than or equal to 10% while for moderate and severe the range is proposed as 11% to 24% and greater than equal to 25%. Based on the variation (for moderate and severe), relevant precautionary measures are developed and implemented to reframe the watershed condition to mild. "The aim of any research is not to find the problem but the solution." With this phrase, the research proposes possible solutions to mitigate the impact of climate change if the vulnerability of the watershed turns out to be moderate or severe. This study develops the simulation by enhancing the green cover of the watershed and running the framework again from stage-6 until the condition of the watershed turns mild. The algorithm is programmed in such a way that the model continues to simulate itself by trying different mitigation measures until the condition of the watershed becomes mild.

The conclusion of the study in the concise and generalized form can be presented as there exists unavoidable evidence of the climate change impact on the Hathmati River which happens to be the important tributary of western India. Lack of mitigating measures could lead to the inevitable consequences which might affect the natural resources and economy of the nation.

The proposed framework could be applied to any watershed with the objective to assess the impact of climate change on the hydrology of the watershed and mitigating the impact by adopting necessary measures.

Data Availability Statement

The thematic maps, namely contour, drainage, land use/land cover, and soil maps have been generated from the CARTOSAT-I, SRTM, BHUVAN, and IRS~ID LISS III satellite data and prepared in ArcGIS on a 1:12500 scale. The weather data have been obtained from the state weather data center and IMD. The geographical & hydro-meteorological information of the study area is taken from the state water resources department.

Conflicts of Interests

All authors have participated in (a) conception and design, or analysis and interpretation of the data; (b) drafting the article or revising it critically for important intellectual content; and (c) approval of the final version. This manuscript has not been submitted to, nor is under review at, another journal or other publishing venue. The authors have no affiliation with any organization with a direct or indirect financial interest in the subject matter discussed in the manuscript.

References

[1] Alexander, L. V., Zhang, X., Peterson, T. C., Caesar, J., Gleason, B., Klein Tank, A. M. G., Vazquez-Aguirre, J. L. (2006). Global observed changes in daily climate extremes of temperature and precipitation. *Journal of Geophysical Research,* 111(D5). doi:10.1029/2005jd006290.

[2] Vose, R. S., Easterling, D. R., & Gleason, B. (2005). Maximum and minimum temperature trends for the globe: An update through 2004. Geophysical Research Letters, 32(23). doi:10.1029/2005gl024379Neitsch G J., Arnold S. L, Kiniry J. R. and Williams, J. R. (2001). *Soil and Water Assessment Tool Theoretical Documentation: Version 2000.*

[3] Chandole V. et al. (2019). Spatio-temporal trend detection of hydro-meteorological parameters for climate change assessment in Lower Tapi river basin of Gujarat state, India, 2019, *Journal of Atmospheric and Solar-Terrestrial Physics*, 195: 105130. https://doi.org/10.1016/j.jastp.2019.105130.

[4] Oza, Markand & Kishtawal, Chandra. (2014). Spatial analysis of Indian summer monsoon rainfall. *Journal of Geomatics.* 8. 40-47.

[5] Gleick, P. H.: 1986, Methods for evaluating the regional hydrologic impacts of global climatic change. *Journal of Hydrology* 88, 97–116.

[6] Burn, D. H.: 1994, Hydrologic effects of climatic change in the West-Central Canada. *Journal of Hydrology* 160, 53–70.

[7] Abeysingha, Nimal & Singh, Man & Sehgal, Vinay & Khanna, Manoj & Pathak, Dr Surendra & Jayakody, Priyantha & Srinivasan, Raghavan. (2015). Assessment of water yield and evapotranspiration over 1985 to 2010 in the Gomti River basin in India using the SWAT model. *Current Science.* 108. 25-2015.

[8] Grillakis, M. G., Koutroulis, A. G., Komma, J., Tsanis, I. K., Wagner, W., Blöschl, G., 2016. Initial soil moisture effects on flash flood generation −a comparison between basins of contrasting hydro-climatic conditions. *J. Hydrol.* 541, 206–217, http://dx.doi.org/10.1016/j.jhydrol.2016.03.007.

[9] Jiang, T., Chen, D. Y., Xu, C., Chen, X., Chen, X., Singh, V. P., 2007. Comparison of hydrological impacts of climate change simulated by six hydrological models in the Dongjiang Basin, South China. *J. Hydrol.* 336, 316–333, http://dx.doi.org/10.1016/j.jhydrol.2007.01.010.

[10] Modrick, T. M., Georgakakos, K. P., 2015. The character and causes of flash flood occurrence changes in mountainous small basins of Southern California under

projected climatic change. *J. Hydrol.: Reg. Stud.* 3, 312–336, http://dx.doi.org/10.1016/j.ejrh.2015.02.003.

[11] Piras, M., Mascaro, G., Deidda, R., Vivoni, E. R., 2016. Science of the Total Environment Impacts of climate change on precipitation and discharge extremes through the use of statistical downscaling approaches in a Mediterranean basin. *Sci. Total Environ.* 543, 952–964,http://dx.doi.org/10.1016/j.scitotenv.2015.06.088.

[12] Sonali, P. and D. Nagesh Kumar, (2013). Review of trend detection methods and their application to detect temperature changes in India, *Journal of Hydrology,* Elsevier, Vol. 476, pp. 212-227.

[13] Arora, Manohar & Singh, Pratap. (2005). Evaluation of temperature trends over India. *Hydrological Sciences Journal.* 50. 10.1623/hysj.50.1.81.56330.

[14] Goswami N. N. et al. 2006. Increasing trend of extreme rain events over India in a warning Environment, *Sciencemag* v.314, pp. 1442-1444.

[15] Dave H. K. et al. (2012). Study of rainfall variation pattern in Hathmati Basin. *Indian Journal of Research*, 1(10). ISSN 2250-1991, v1 pp 70-71.

[16] Yu P. et al. Effects of climate change on evapotranspiration from paddy fields in Southern Taiwan. (2002). *Climatic Change,* v 54, pp 165–179. https://doi.org/10.1023/A:1015764831165.

[17] Rupa, K. K. et al. (2002). On forecasting the Indian summer monsoon: the intriguing season of 2002, *Current Science,* v 83, issue 4, pp 394-403.

[18] Basu S. et al. (2018). Response of grassland ecosystem to monsoonal precipitation variability during the Mid-Late Holocene: Inferences based on molecular isotopic records from Banni grassland, Western India. *PLOS ONE*, 14(4). https://doi.org/10.1371/ journal.pone.0212743.

[19] Sirois, F. and Pychyl, T. (2013). Procrastination and the priority of short-term mood regulation: consequences for future self. *Social and Personality Psychology Compass,* 7 (2). 115 - 127. ISSN 1751-9004 https://doi.org/10.1111/spc3.12011.

[20] Wang, Xueqin & Yu, Qiqing. (2005). Unbiasedness of the Theil–Sen estimator. Journal of nonparametric statistics. *J Nonparametr Stat.* 17. 685-695.10.1080/10485250500039452.

[21] Beven, K. J. (2000). *Rainfall-Runoff Modeling: The Primer.* Wiley: Chichester, England. DOI:10.1002/9781119951001.

[22] Singh, V. P. and Woolhiser, D. A. (2002). Mathematical modeling of watershed hydrology. *Journal of Hydrologic Engineering,* 7(4), 270-292.

[23] Neitsch G J., Arnold S. L, Kiniry J. R. and Williams, J. R. (2001). *Soil and Water Assessment Tool Theoretical Documentation.* Version 2000.

[24] Jayakrishnan, R. et al. (2005). Advances in the application of the SWAT model for water resources management, 2005, *Hydrological Processes*, 19. 10.1002/hyp.5624.

[25] Shafiq M. et al. Assessment of present and future climate change over Kashmir Himalayas, India. https://doi.org/10.1007/s00704-019-02807-x.

[26] Narsimlu, Boini & Gosain, Ashvani & Chahar, Bhagu & Singh, Sudhir & Srivastava, Prashant. (2015). SWAT model calibration and uncertainty analysis for streamflow prediction in the Kunwari River Basin, India, using sequential uncertainty fitting. *Environ. Process.* 2. 10.1007/s40710-015-0064-8.

[27] Abbaspour K, Vejdani M, Haghighat S (2007) SWAT-CUP calibration and uncertainty programs for SWAT. In: *MODSIM 2007 International Congress on Modelling and Simulation, Modelling and Simulation Society of Australia and New Zealand, 2007.*

[28] Duan Q., Schaake J., Andreassian V., Franks S., Goteti G., Gupta H. V., Gusev Y. M., Habets F., Hall A., Hay L., Hogue T., Huang M., Leavesley G., Liang X., Nasonova O. N., Noilhan J., Oudin L., Sorooshian S., Wagener T., Wood E. F. (2006) Model parameter estimation experiment (MOPEX): an overview of science strategy and major results from the second and third workshops. *J Hydrol* 320:3–17.

[29] Krishnan P. et al. (2019). Framework for mapping the drivers of coastal vulnerability and spatial decision making for climate-change adaptation: A case study from Maharashtra, India. *Ambio*, 48: 192–212. https://doi.org/10.1007/s13280-018-1061-8.

[30] Jaiswal, R. K., Yadav, R. N., Lohani, A. K. et al. Water balance modeling of Tandula (India) reservoir catchment using SWAT. *Arab J Geosci,* 13, 148 (2020). https://doi.org/10.1007/s12517-020-5092-7.

[31] Ghosh, S., Das, D., Kao, S. C. and Ganguly, A. R., 2012. Lack of uniform trends but increasing spatial variability in observed Indian rainfall extremes. *Nature Climate Change,* 2(2), p.86.

[32] IPCC, (2001b). *Climate Change Impacts, Adaptations and Vulnerability* (ed. by J. J. McCarthy., O. F. Canziani., N. A. Leary., D. J. Dokken & K. S. White), Cambridge University Press, Cambridge, UK.

[33] IPCC., (2007). *Fourth Assessment Report (AR4).* http://www.ipcc.ch/publications_and_data/publications_ipcc_fourth_assessment_report_synthesis_report.

[34] Muhammed Anawar Saadat and Saiful Islam, A. K. M., (2011). Impact of climate change on rural livelihoods: case study. *3rd International Conference on Water & Flood Management (ICWFM-2011),* http://teacher.buet.ac.bd/ak msaifulislam/publication/ICWFM2011fullpaper137.pdf.

[35] Zubair., and Ayodeji Opeyemi., (2006). *Change Detection in Land Use/ Land Cover using Remote Sensing Data and GIS, (A Case Study of Ilorin and its Environs in KwaraState).* M.Sc., Project, http://www.geospatialworld.net/uploads/.../OpeyemiZubairThesis DOC.doc.

Chapter 7

Groundwater Vulnerability Modeling

Osama Gazal[1,*], Saeid Eslamian[2], Hazim Elnaser[3] and Jihad Mahamid[4]

[1]Climate Change and Environmental Studies Department, Ministry of Water & Irrigation, Amman, Jordan
[2] Department of Water Engineering, College of Agriculture, Isfahan University of Technology, Isfahan, Iran
[3]Middle East Water Forum, Al Fuheis, Jordan
[4]Ministry of Water and Irrigation Secretary General, Jordan

Abstract

Groundwater sources are very important, particularly in dry areas, and the sustainability of these resources necessitates serious efforts to protect them from contamination. Digital models have become the mainstay of risk assessment studies and groundwater aquifer protection through intrinsic or specific contamination susceptibility models. Groundwater is more likely to harm the consumer population if exposure occurs since fewer measures are also taken to eradicate the contaminant before it reaches the public. For this reason, the conservation of groundwater is of considerable importance, particularly in arid areas with highly vulnerable karstic and fractured basaltic aquifers. Recently, mapping the vulnerability of groundwater contamination has become a key step in groundwater risk assessment and protecting groundwater quality.

This chapter consists of two parts the first discusses the components of the groundwater risk assessment and examines the evolution of the groundwater vulnerability modeling. The scope of this review involves the design of standard methods for assessing contamination potential, as well as the development of a system that can be easily displayed on maps.

* Corresponding Author's Email: osama_gazal@mwi.gov.jo.

In: Handbook of Uncertainty in Eurasian Forecasting (HEF)
Editors: Saeid Eslamian, Mir Bintul Huda and Faezeh Eslamian
ISBN: 978-1-68507-966-6

Thus, this review is not intended to evaluate the various kinds of groundwater vulnerability modeling. Rather, it is meant to serve as a general base for the variety and evolution of vulnerability evaluation methodologies. The second part explains the visualization of groundwater vulnerability by a systematic study carried out to establish an effective method for assessing the groundwater risk of renewable aquifers in the arid agricultural area by developing a multipurpose database in the GIS setting and validating the overlaying vulnerability method developed by comparing the findings with the groundwater quality characteristics of the area and with former visualization.

In this chapter, the authors conduct a review of the existing literature on groundwater vulnerability assessment before describing an improved methodology based on the combination of DRASTIC methods and simple modeling of nitrate contaminants from the land surface to groundwater.

Keywords: groundwater, risk assessment, vulnerability models, contamination, GIS, intrinsic vulnerability, specific vulnerability, overlaying model

Part I

Introduction

John F. Kennedy said, "*Anyone who can solve the problems of water will be worthy of two Nobel prizes - one for peace and one for science.*" Rapid industrial and agricultural expansion today has the potential to contaminate both surface and underground water resources. However, groundwater is treated even less than surface water because it seems to be a safer source. This also suggests that it is more likely to have a negative effect on the consumer population if exposure occurs since fewer measures are also taken to eradicate the contaminant before it reaches the public. Parallel to this rising risk, numerous countries have constructed groundwater vulnerability maps for this purpose over the previous few decades. One of the goals of such maps is to show decision-makers the regions where there is a high danger of contaminants reaching groundwater under hydrogeological conditions. All activities that potentially contaminate groundwater resources should be prohibited in these extremely sensitive areas. On such maps, areas with the higher natural protection of groundwater against pollution are also depicted. Current regulations must be enforced, and new regulatory measures and legislation must be enacted to safeguard groundwater aquifers and limit the use of fertilizers. A reduction in the amount of groundwater used for irrigation, as well as the creation of land-use strategies, will be required.

The conservation of groundwater is of considerable importance, particularly in highly vulnerable karstic, fractured basaltic, and sandstone aquifers (Gazal 2021, Gazal & Eslamian 2021a). Protecting groundwater from contamination is essential for safe, sustainable drinking water management (Sasakova et al. 2018), and it is not a new activity, but since ancient times many cultures have considered the importance of protecting the quality of the groundwater and ensuring that it is not contaminated. Foster and Loucks (2006) stated the old Arabic Proverb "*Into the well from which you drink do not throw stones.*" Groundwater represents about 30% of the planet's freshwater, while only 0.3% is in form of surface water bodies (lakes and rivers) and more than 68% of the planet's freshwater is in form of glacial (Yano et al. 2015). Groundwater accounts for 98% of all available freshwater on earth (Velis et al. 2017). Total reservoirs of fresh groundwater are around 100 times bigger than reservoirs of fresh surface water (Fitts 2013), see the modified sketch after Gleick (1996) (Figure 1).

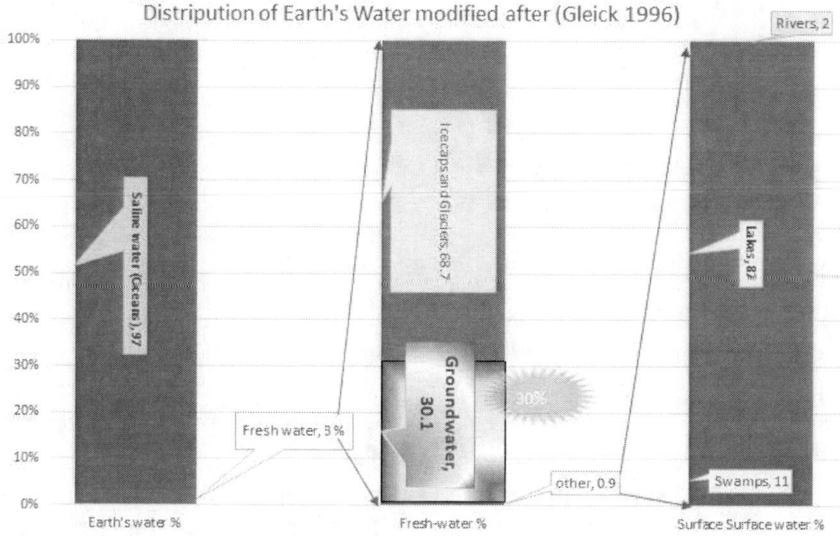

Figure 1. Distribution of Earth's water modified after Gleick (1996).

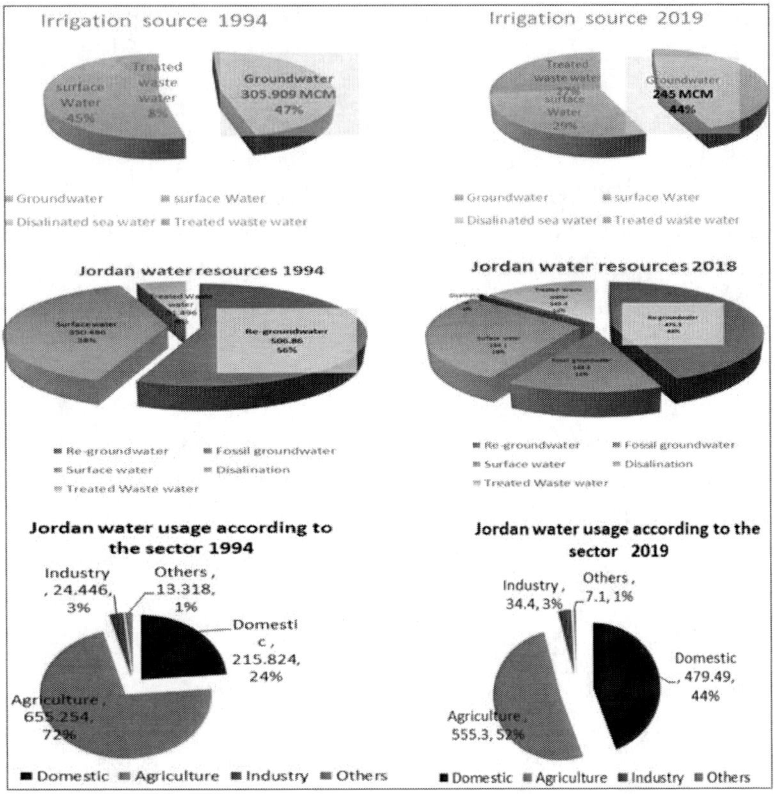

Figure 2. Jordan water resources, usages, and irrigation water resources in (MCM/year), for the years 1994 and 2019, own processing and the raw water data retrieved from WIS (2021).

Currently, in Jordan groundwater resources cover 44% of the irrigation requirements and more than 58% of the country's water requirements (Figure 2). The percentage of the water used in agriculture to the total water uses in 2019 was 51% while it was above 70% at the beginning of the nineties. This is due to the agricultural policies, increasing the profitability and efficiency of the irrigation methods to cope with the water scarcity in Jordan (WIS 2021). Besides, groundwater is, in fact, the world's largest extracted natural material (Vogelbacher et al. 2019). This increase is significant in arid and semi-arid areas where groundwater can be the only available drinking water resource (MWI, 2004; Margane & Hobler 1994). The low-efficiency uses of water resources in the dry and semi-dry areas agitate the problems of losing these resources, while several factors contribute to the deterioration of groundwater aquifers, exacerbated by drought and climate change impacts affecting arid and semi-arid areas that threaten groundwater resources (El-Beltagy & Madkour 2012; Balon & Dehnad 2006).

Groundwater resources contain 99% of all freshwater accessible on the earth and are widely distributed across all continents (Puri 2009). They can be classified as renewable (phreatic aquifer) or un-renewable (confined aquifer), according to the natural flow and storage of (GW) aquifer (Ge & Gorelick 2015), where the renewal periods range from less than 10 years up to 100,000 years. However, both classes can be considered non-renewable according to the limited possibility of natural replenishment due to low annual rainfall and over-abstraction in many arid and semi-arid areas (Margat et al. 2006). The abstraction of groundwater dates to the period of early human existence (Ajami 2020). However, the use of groundwater has increased since the second half of the last century of the 2nd millennium, the 20th century, due to major developments in well-drilling and pumping technology (Foster et al. 2013).

Ground water differs from surface water as it is naturally protected from contamination caused by human agricultural and industrial activities on the land, whereas this natural protection varies from region to region, depending on the type and load of contaminant and on the ability of the above-water layers to prevent the downflow of pollutants (Morris & Foster 2001). Agricultural contaminants have the most severe and widespread effects on groundwater and the most studied contamination parameter is a nitrate which mainly increases where there are intensive agricultural activities. Agricultural activities can result in high concentrations of nitrate in the groundwater (USGS 2017). Groundwater quality degradation and low annual rainfall may create other environmental problems too, while some of the pollution impacts in the groundwater basin might be irreversible such as loss of biodiversity and deterioration of human health (IdRC & IUCN 2006).

Groundwater Sustainability

Groundwater is an important resource for the ecosystem and one that needs to be properly managed, making the world more focused on sustainable groundwater resource management studies that continue to increase and improved (Kresic & Stevanovic 2009). Preserving continuous and adequate supplies of drinkable water from groundwater sources is a critical issue for sustained agriculture, industry, and domestic use throughout arid and semi-arid areas, specifically in extremely developing water-scarce countries that suffer from huge refugee fluxes from the surrounding countries like Jordan the case study of this research.

Contaminants can reach groundwater via transport through the soil profile and, depending entirely on their nature, can have very serious consequences, making it a key requirement for sustainable development to protect groundwater resources from contamination that will increase the sustainability of high-quality drinking water (Sasakova et al. 2018).

The definition of sustainable development according to Brundtland (1987) is a *"development that meets the needs of the present without compromising the ability of future generations to meet their own needs."* The concept of groundwater sustainability was defined by Alley et al. (1999) as the *"development and use of groundwater in a manner that can be maintained for an indefinite time without causing unacceptable environmental, economic, or social consequences."*

Two main factors are controlling the sustainability of the groundwater recourse, the climate change affects that will cause the expansion of the arid and semi-arid regions and decrease water availabilities (Balon & Dehnad 2006). The second factor is demographic growth and the rapid economic activities development, combined with the agricultural intensification of food production, which is threatening to expand the arid areas and innervating the natural resources (Fernández-Cirelli et al. 2009). Water security in arid and semi-arid regions requires sustainable and integrated management tools and a clear understanding of the interactions between the groundwater aquifers and human surface activities (Dawoud & Sallam 2012).

Sustainable groundwater management is dictated not only by the availability of groundwater resources but also by water quality degradation (Foster et al. 2013), and the process of studying groundwater quality has progressed considerably over the last decade with several significant advances in groundwater investigation technology including enhanced integration of groundwater statistical models and contaminant transport with external systems.

Groundwater Contamination Risk Assessment

Groundwater risk assessment (GRA) is defined by Morris & Foster (2001) "as the probability that groundwater in the aquifer will become contaminated to an unacceptable level by activities on the immediately-overlying land-surface, the risk will be the result of the interaction between the subsurface contaminant load and the aquifer pollution vulnerability at the location concerned." The identification of vulnerable groundwater levels is crucial to the sustainable management and conservation of limited groundwater resources. Arid and semi-arid areas are specifically reliant on groundwater sources. However, due to several factors linked to the disproportionate use of groundwater and agricultural activities, contamination has increasingly become a serious problem, specifically in areas where fertilizers are extensively used to enhance the productivity of a soil characterized by low fertility. The common causes of groundwater quality deterioration can be classified by genesis explained by Foster et al. (2002). This classification assumed four classes of groundwater quality problems aquifer pollution, wellhead contamination, saline water intrusion, and naturally occurring contaminants.

The present work is designed to address the first problem the "aquifer contamination" in a comprehensive groundwater contamination risk assessment (GRA) of a renewable groundwater aquifer in a dry agricultural area. The aquifer contamination has the following underlying causes according to Foster et al. (2002) "inadequate protection of vulnerable aquifers against manmade discharges and leachates from urban/industrial activities and intensification of agricultural cultivation" and the following contaminants of concern "pathogens, nitrate or ammonium, chloride, sulfate, boron, arsenic, heavy metals, dissolved organic carbon, aromatic and halogenated hydrocarbons, certain pesticides." The time needed for the surface contaminants to reach the groundwater aquifer depends on several factors related to the load of the contamination and the characteristic of the aquifer. While the contamination load mainly depends on the land use activities (Figure 3), this explains the significance of contracting aquifer pollution vulnerability and the difference between the contamination load according to the land uses. Groundwater resources are safer and less vulnerable to pollution than the surface water resources (such as rivers and lakes) since groundwater aquifers are naturally protected by soil and unsaturated layers over the water table. But that does not mean that groundwater contamination is less serious than contamination of surface water. Groundwater is not fully out of harm's way just because it is out of sight, and just because a contaminated aquifer might not be as offensive to our senses as a polluted river, that does not mean that groundwater contamination is less serious than surface water contamination (Freeze and Cherry 1979; Foster 2001).

Groundwater contamination is more difficult to detect and track and the source is more difficult to locate than surface water. Maybe most important, if the aquifer is contaminated, the job of cleaning it up, even if it is feasible, is immense. Perhaps more than in the case of surface water contamination, the maxim of groundwater contamination is that prevention is better than treatment (Price 2013). However, the potential for this natural protection depends on a variety of factors, and natural protection varies from place to place, depending on the hydrogeological nature of the aquifer and the contamination load on the surface.

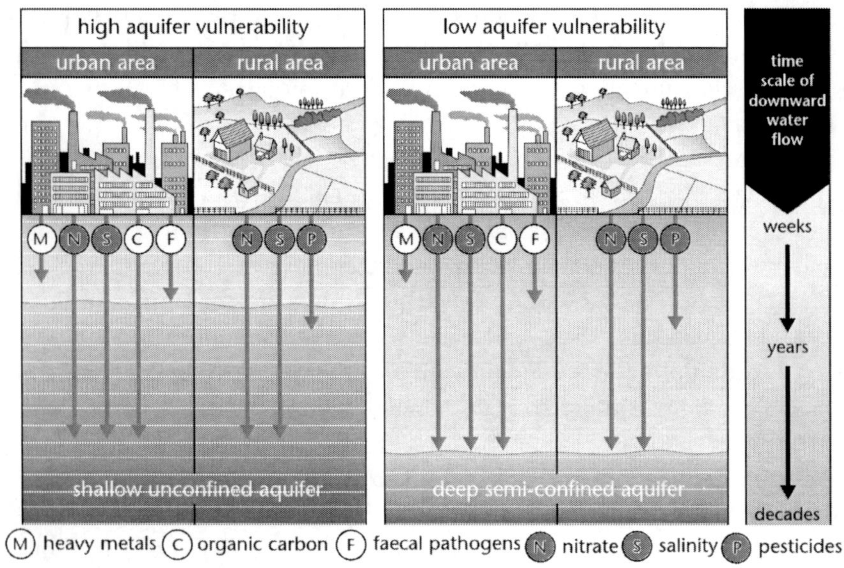

Figure 3. Significance of contrasting aquifer pollution vulnerability (Foster et al. 2002).

Groundwater aquifer contamination occurs where the contaminant load produced by man-made discharges and leachates (from urban, manufacturing, agricultural, and mining activities) is poorly regulated and exceeds the natural attenuation ability of the overlying soils and the unsaturated strata. Figure 4, and Table 1 summarize the common groundwater contaminations and associated pollution sources. The biochemical degradation and chemical reaction are the product of the auto-elimination of pollutants during sub-surface transport in the vadose (unsaturated) zone, but contaminant retardation processes due to sorption phenomena are also of significance as they increase the time available for processes resulting in contaminant elimination. Not all subsoil profiles and underlying unsaturation layers, however, are equally successful in attenuating pollutants, and aquifers would be especially vulnerable to contamination where, for instance, strongly fissured consolidated rocks are present.

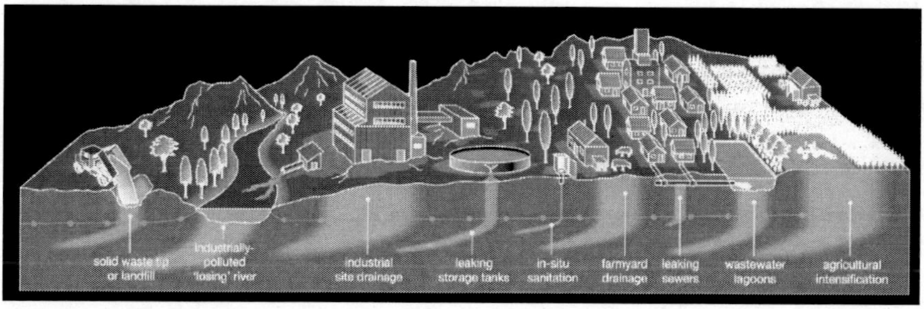

Figure 4. Common processes of groundwater pollution (Foster et al. 2002).

Table 1. Common groundwater contaminants and associated pollution sources (Foster et al. 2002)

Pollution source	Type of contaminant
Agricultural activity	Nitrates; ammonium; pesticides; fecal organisms
In-situ Sanitation	nitrates; halogenated hydrocarbons; microorganisms
Gas Stations and Garages	Aromatic hydrocarbon; benzene; phenols; halogenated hydrocarbons
Solid Waste Disposal	Ammonium; salinity; halogenated hydrocarbons; heavy metals
Metal Industries	Trichloroethylene; tetrachloroethylene; halogenated hydrocarbons; phenols; heavy metals; cyanide
Painting and Enamel Works	Alkylbenzene; halogenated hydrocarbons; metals; aromatic hydrocarbons; tetrachloroethylene
Timber Industry	Pentachlorophenol; aromatic hydrocarbons; halogenated hydrocarbons
Dry Cleaning	Trichloroethylene; tetrachloroethylene
Pesticide Manufacture	Halogenated hydrocarbons; phenols; arsenic
Sewage Sludge Disposal	Nitrates; halogenated hydrocarbons; lead; zinc chromium;
Leather Tanneries	Chromium; halogenated hydrocarbons; phenols
Oil and Gas Exploration/Extraction	Salinity (sodium chloride); aromatic hydrocarbons
Metalliferous and Coal Mining	Acidity; various heavy metals; iron; sulfates

The degree of attenuation in any given setting will also vary widely with pollutant forms and pollutant processes. Natural soil profiles actively attenuate many, but not all, water pollutants (Glower 1983; Foster et al. 2002) briefly clarified the key mechanism of attenuation of pollution in the groundwater system as follows:

1. Many, but not all, are subject to naturally occurring attenuation processes when waterborne contaminants move below the ground surface through the soil layer. This can be a potentially efficient device if the contaminant enters through a natural soil profile because soils are typically aerated by soil fauna and are very biologically active. This favors bacterially mediated degradation, which in an aerobic environment is more successful for many compounds.
2. Soils can also have elevated content of clay and/or humus, creating opportunities for sorption and ion exchange.
3. However, in an urban environment, by systems that bypass the soil zone, such as septic tanks, and latrines, many pollutants directly penetrate the subsurface. As these systems are built to dispose of waterborne wastes rapidly, they will almost invariably enforce a higher hydraulic loading, or surcharge, than contaminants entering through the soil layer. The increased speeds of percolation resulted in a subsequent decrease in residence times and the possibility of contaminant degradation.
4. Contaminant attenuation processes in the unsaturated zone continue to a lesser degree with depth. Figure 5 qualitatively shows their relative value in the soil, above, at, and below the groundwater table.

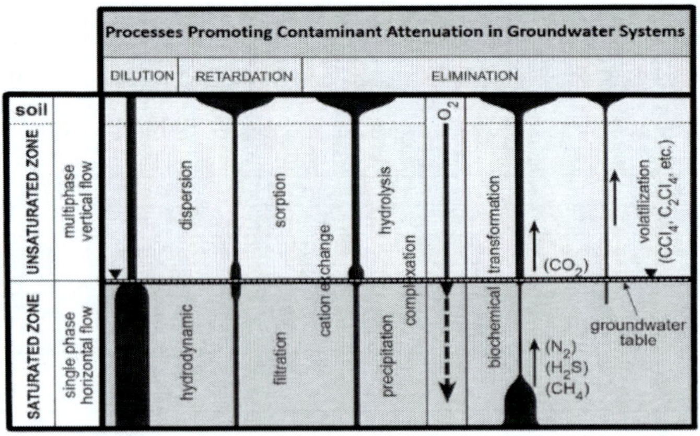

Figure 5. The main process promoting contamination attenuation in groundwater systems (Glower 1983).

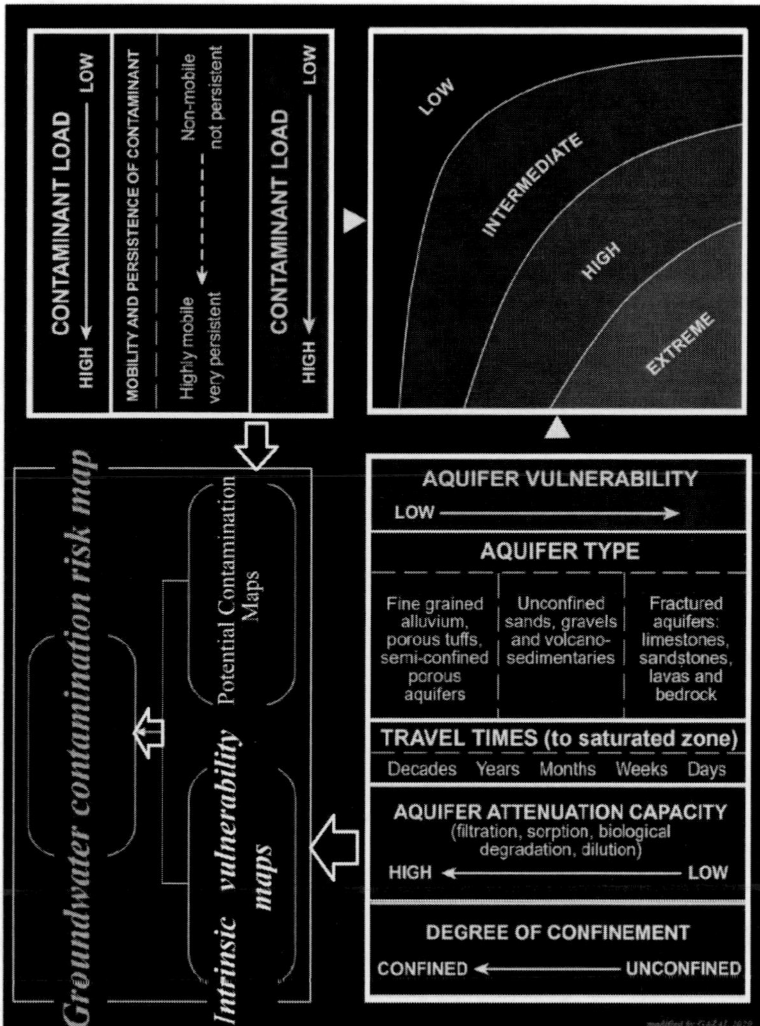

Figure 6. The conceptual basis scheme for groundwater contamination risk (Foster et al. 2002; Foster & Hirata 1988; Civita, 1990).

The groundwater contamination risk could then be described as the possibility that groundwater in the aquifer would become polluted by activities on the immediately overlying land surface to an unacceptable degree. The danger would arise from the relationship between the load of the subsurface contaminant and the vulnerability of aquifer contamination at the location concerned. Then the groundwater contamination risk can be evaluated by studying two components, the contamination load, and the natural groundwater potential protectiveness. The most logical approach in the groundwater contamination risk assessment is studying the result of the interaction between the subsurface contaminant load and the aquifer pollution vulnerability at the location concerned (Foster et al. 2002; Foster & Hirata 1988; Civita 1990). Figure 6 explains the general scientific concepts in this regard. Implementing such a simple scheme we may have high groundwater vulnerability, but there is no chance of contamination in the absence of a substantial contaminant load, and vice versa. Both are perfectly consistent.

In addition, the contaminant load may be managed or changed, but the vulnerability of the aquifer is an inherent, natural aspect (other than where the geological profile is removed or disrupted by quarrying, mining, or civil engineering excavation). Therefore, the groundwater contamination risk map created in this study with a simple modified overlying approach (a combination between a modified DRASTIC model and land use) is worthwhile to be used and apply the same technique to construct a groundwater contamination risk map for the whole country (Jordan) and any related dry areas in the world. However, it is more important to rely on the vulnerability map itself to manage the land use process because the contamination status may change over time, but the natural groundwater potential protectiveness is fixed over time unless the geological profile is removed or disturbed by humans' activities. To conclude and according to NRC (1993) the assessment of groundwater vulnerability can be done in two ways:

1. The Groundwater Risk Assessment of specific vulnerability is a combination of the intrinsic vulnerability and the potential or the actual sources of contamination.
2. The Assessment of the General Groundwater Aquifer intrinsic vulnerability: (In the modeling part of this study DRASTIC model was used to assess the intrinsic vulnerability through several modified approaches). The intrinsic (inherent) vulnerability depends only on the characteristics of the groundwater aquifer, which are as follows (Napolitano 1995):
    a) Hydrogeology of the unsaturated layers (vadose zone): It defines the vertical and horizontal permeability and, consequently, controls the speed of pollutant diffusion and the capability of the rocks to attenuate the action of the pollutant;

b) Water table depth: It defines the thickness of the vadose zone (from the topographic surface) and is proportional to the capability of attenuation from pollutants;
c) Hydro-lithologic characteristics of the aquifer: It controls the diffusion of the pollutants when they reach the saturated zone.

A groundwater risk map generally, is a combination of existing hazardous substances and the aquifer vulnerability. But due to the difficulties of recognizing all possible sources of contaminants and the presence of a real contamination problem, the susceptibility is more directed towards a particular contaminant which is in the case study of this dissertation the nitrate contaminant.

Some Terms Related to the Groundwater Vulnerability

1. The "Aquifer vulnerability" is defined by Albinet & Margat (1970) as "the possibility of percolation and diffusion of contaminants from the ground surface into natural water table reservoirs, under natural conditions."
2. The vulnerability was described by Villumsen et al. (1983) as "the risk of the chemical substance used or disposed of on or near the ground surface to influence groundwater quality."
3. The vulnerability was described by Vrba & Zaporožec (1994) as "an intrinsic property of a groundwater system that depends on the sensitivity of that system to human and/or natural impacts." Besides they defined the groundwater vulnerability relates to subsurface contamination from nonpoint pollution as the measure of the "degree of insulation" of groundwater from the land surface contaminants."
4. Groundwater vulnerability was described by Bachmat & Collin (1987) as "the sensitivity of groundwater quality to anthropogenic activities which may prove detrimental to the present and/ or intended usage or value of the resource."
5. Aquifer pollution vulnerability was described by Foster (1987) as "the intrinsic characteristic which determines the sensitivity of various parts of an aquifer to being adversely affected by an imposed contaminant load." And described the "Groundwater Pollution Risk" as "the interaction between (a) the natural vulnerability of the aquifer,

and (b) the pollution loading that is, or will be applied on the subsurface environment because of human activity."

6. The vulnerability of a Hydrogeological system was explained by Sotornikova & Vrba (1987) as "the ability of this system to cope with external, natural and anthropogenic impacts that affect its state and character in time and space."
7. The United States General Accounting Office (USGAO) (1991) described "Hydrogeological Vulnerability" as "a function of geologic factors, as soil texture and depth to groundwater." While the "Total Vulnerability" is "A function of these hydrogeologic factors as well as the pesticide use factors that influence the site's susceptibility," and "Total Risk" as "this last approach is even broader, for it incorporates the size of the population at risk from potential pesticide contamination- that is, the number of people who obtain their drinking water from groundwater in the area."
8. Aquifer Vulnerability was described by Pettyjohn et al. (1991) as "the geology of the physical system determines vulnerability," and "Aquifer Sensitivity" as "aquifer sensitivity is related to the potential for contamination, i.e., an aquifer that has a high degree of vulnerability and is in areas of high population density, is the most sensitive."
9. The United States Environmental Protection Agency (USEPA) (1993) described "Aquifer Sensitivity" as "the relative ease with which a contaminant (in this case a pesticide) applied on or near the land surface can migrate to the aquifer of interest. Aquifer sensitivity is a function of the intrinsic characteristics of the geologic materials of interest, any overlying saturated materials, and the overlying unsaturated zone. Sensitivity is not dependent on agronomic practices or pesticide characteristics," and the "Ground Water Vulnerability" is "the relative ease with which a contaminant (in this case a pesticide) applied on or near the land surface can migrate to the aquifer of interest under a given set of agronomic management practices, pesticide characteristics, and hydrogeologic sensitivity conditions."
10. The National Research Council (NRC) (1993); (NRC) (1994) described the "Groundwater Vulnerability to Contamination" as "the tendency or likelihood for contamination to reach a specified position in the groundwater system after introduction at some location above the uppermost aquifer."

Review of the Groundwater Vulnerability Mapping Approaches

Scientists and resource managers have tried to establish a methodology for estimating areas that are more likely to become polluted by activities on or near the surface of the earth than others (NRC 1993). Creating a vulnerability map for the groundwater aquifers is a relatively modern procedure in groundwater sustainability studies (Vrba and Zaporozec 1994). Groundwater Vulnerability is one of the terms that led to a breakthrough in water resource sustainable management, enabling decision-makers and researchers in the field of groundwater pollution to identify the causes of pollution and the severity of pollution from one place to another and help to put the guidelines to create groundwater protection zones. Years before the notion of groundwater vulnerability and protection emerged in the 1960s, along with the rapid expansion and application of zoning and indexing-rating models, most studies employed coherent physicochemical parameters for geological and hydrogeological characteristics. For example, the travel time needed for the pollution front to enter groundwater from the source of contamination, the retardation factor, the hydraulic resistance of the cover, and water-bearing deposits, the contamination front to reach groundwater from the source, the retardation factor (ratio of velocities of seepage water, and contaminant particles) (Faybishenko et al. 2013).

The first mention of the concept of groundwater vulnerability as an alternative to "natural protection against contamination" in the 1960s was made by French hydrogeologist Jean Margat, currently, Vice-President of the Association du Plan Bleu and regularly consulted as an expert by international organizations such as FAO, UNESCO, World Bank, and UNDP, and won the IASH, UNESCO and WMO International Hydrology Prize in 2008. Margat (1968) for the first time defined the concept of groundwater vulnerability as "*the ease with which the contaminant introduced into the ground surface may enter and diffuse into groundwater*" and this simply, the lower the natural protection, the higher the vulnerability (Nguyet & Goldscheider 2006). And after the first definition by Dr. Margat in the 1960s, this basic concept of groundwater vulnerability has taken on a range of definitions in the technical literature, and several terminologies have since been given.

The National Research Council (NRC) (1993) identified two general types of vulnerability assessments. The first type is needed to address the specific vulnerability of the site to a specific contaminant or contaminant class due to human activity. The second type is intended to address the intrinsic vulnerability of groundwater without consideration of the attributes and behavior of specific contaminants. While according to the NRC (1993) the approaches to investigate groundwater vulnerability can be categorized into three main traditional approaches (Table 2):

Table 2. Selected methods used to evaluate groundwater vulnerability (NRC 1993)

Method	Reference	Map Scale*	Reference Location	Intrinsic and/ or Specific
Overlay and Index methods				
Kansas Leachability Index	(Kissel et al. 1982)	Small	Soil	Intrinsic
DRASTIC	(Aller et al. 1987)	Variable	Groundwater	Intrinsic
California Hotspots	(Cohen et al. 1986)	Large	Water Table	Intrinsic and Specific
Washington Map Overlay Vulnerability	(Sacha et al. 1987)	Small	Groundwater	Intrinsic and Specific
SEEPAGE	(Moore 1988)	Variable	Groundwater	Intrinsic
Lowa Groundwater Vulnerability	(Hoyer & Hallberg 1991)	Small	Groundwater	Intrinsic
EPA/UIC	(Pettyjohn et al. 1991)	Small	Groundwater	Intrinsic
Process-Based Simulation models				
PESTANS	(Enfield et al. 1982)	Large	Soil	Specific
BAM	(Jury et al. 1983)	Large	Soil	Specific
MOUSE	(Steenhuis et al. 1987)	Large	Groundwater	Specific
PRZM	(Carsel et al. 1984)	Large	Soil	Specific
RF/AF	(Rao et al. 1985)	variable	Soil	Specific
GLEAMS	(Leonardo et al. 1987)	Large	Soil	Specific
CMLS	(Nofziger & Hornsby 1986)	Large	Soil	Specific
RITZ/VIP	(McLean et al. 1988)	Large	Soil	Specific
LEACHM	(Wagenet & Huston 1989)	Large	Large	Specific
RUSTIC	(Dean et al. 1989)	Large	Groundwater	Specific and Intrinsic
Statistical methods				
Discriminant Analysis	(Teso et al. 1988)	Small	Groundwater	Specific
Regression Analysis	(Chen & Druliner 1988)	Small	Groundwater	Specific

* "Large Scale" means that the method is typically applied at a level of detail of at least a 1:24,000 scale map to a small spatial area. "Small Scale" means that the method is typically applied at a level of detail less than that of a 1:50,000 scale map to a larger spatial area (NRC 1993).

1. Overlay and index methods: Involves obtaining and combining maps of the parameters that affect the transport of contaminants from the surface to groundwater, then assigning an index value to those

parameters; the results are a spatially oriented vulnerability index. The combination of several parameters by assigning a numerical index or score to each parameter attribute with values indicating the degree to which that parameter indicates the possibility of protection of groundwater from contamination, or otherwise, in the study area. The simplest overlay approach (schemes) defines areas where parameters that imply vulnerability have the same effects, such as shallow groundwater and sandy soil, which would represent similar vulnerability measurements. While, based on many parameters' inputs, more sophisticated approaches assign numerical scores.

2. Process-based simulation models: Involves numerical modeling and is useful at the local level but not the regional level, therefore the map scale in such an approach is small and an example of those models PRZM, GLEAMS, and LEACHM which can predict the fate and transport of contaminants from known sources with notable accuracy in a localized area by applying fundamental physical principals to predict the flow of water in porous media and the behavior of chemical constituents carried by that water. Models in these groups range from indices based on simple transport models to analytical solutions for one-dimensional transport of contaminants through the vadose zone to saturated zones on to several phases (two or three-dimensional models).
3. Statistical methods: Involves correlating actual water quality data to spatial variables and requires a large amount of site-specific data. These models are often used as a validation test for other models.

Although a comprehensive review of the applied groundwater vulnerability assessment methods since the first mention of this term by Margat (1968) may indicate the following list of the most significant groundwater vulnerability and protectability assessment approaches:

1. Hydrogeological zoning methods (HZM): This approach was scientifically first explained by Margat (1968) and used for the assessment of groundwater vulnerability, also it was used by (Vrana 1968; Vrana 1984; Albinet & Margat 1970; Josopait & Schwerdtfeger 1979; Ball et al. 2004). Also, Margane et al. (1997) called this approach the hydrogeological complex and setting methods (HCS) and defined it as a method of assessing vulnerability to groundwater by defining groups of two or more vulnerabilities. Classes are based

on parameters considered to be indicative of the groundwater vulnerability under such hydrogeological conditions. This form of mapping is used mostly for maps of medium to large-scale areas.

2. Mapping vulnerability of the shallow aquifers by estimation of the groundwater age: The main concept behind making vulnerability maps using age data is that the older the groundwater, the less chance it has of being anthropologically polluted (Kazemi et al. 2005).
3. Parametric (scoring and index system) methods:
   a) Matrix systems, assess groundwater vulnerability based on a selection of two or more parameters considered to be representative of a certain area (UNESCO-IHP 2004; Vrba & Zaporožec 1994; Margane et al. 1997).
   b) Rating Systems and Point Count System Models:
      - Rating systems: Using various hydrogeological parameters with defined ranges of ratings according to their variation. The overall rating is determined by overlaying the ratings for these various parameters and then splitting the overall rating into different vulnerability rates. GOD and GLA approaches can be assigned to this method. GOD was developed by Foster (1987); and Foster & Hirata (1988) for geological conditions of Great Britain, the German GLA-Model was developed by Hölting et al. (1995) and used in Jordan groundwater studies by Margane et al. (1997). Also, the SUPRA model was proposed by Zaporozec (1985); Zaporozec (2002), and the German PI model of the regional assessment of intrinsic groundwater vulnerability was created by Goldscheider (2005), and the German COP model by Vías et al. (2006) which was developed through the EU project COST-620 (Zwahlen 2004) and the latest groundwater vulnerability map for Jordan created by applying COP model (BGR & MWI 2018), and the Pesticide root zone model for groundwater (PRZM-GW) to estimate concentrations of pesticides in groundwater for drinking water exposure assessments since 2012 which was created as part of the North American Free Trade Agreement (NAFTA) to develop a unified groundwater modeling protocol (HC and USEPA 2012) developed by the US Environmental protection agency (USEPA), office of pesticide programs (OPP) (HC & USEPA 2012)
      - Point count system models (PCSM): Also called parameter weighting and rating methods which are used to measure the most

important hydrogeological characteristics and use the same approach as rating methods but attribute different weights in the form of a multiplier to reflect the importance of each parameter for the overall assessment of groundwater vulnerability so all the (PCSM) models belong to the rating system. Among the scoring and index approaches the DRASTIC model is one of the earliest methods for determining aquifer vulnerability and the first point count system models (PCSM) or parameter weighting and rating methods. The most widespread PCSM method of evaluation of the groundwater vulnerability developed by Aller et al. (1987) ordinary DRASTIC index can vary within the range of 23–230 (intrinsic vulnerability) or 26–260 for agricultural DRASTIC which used in this study (vulnerability to pesticides). DRASTIC later was subjected to several modifications of its parameters weights by applying statistical methods, also, to be modified by adding other parameters as Denny et al. (2007) proposed to include the structural geology by adding the geological lineaments maps and other index-rating assessment methods were developed like the Italian model SINTACS developed by Civita & Maio De (2004). Unlike DRASTIC, the scores table for the SINTACS indicator includes more detailed lithological variations and disruptions, EPIK model was created especially for use in karst areas in Switzerland by Doerfliger et al. (1999), and the Chines model DRAV was developed by Zhou et al. (2010)

4. Confined groundwater protectability: can be measured using Goldberg's qualitative groundwater protection assessment (Goldberg 1983; Goldberg 1987) of the thickness of the overly permeable (low-permeable) confining layers, including the groundwater hydraulic head ratio in confined and upper unconfined aquifers.
5. Modeling methods:
    a) Deterministic: The deterministic (processed-based) approach involves numerical modeling and is useful at the local level but not the regional level, for example, the deterministic approach to the A3D model was constructed in the work of Loague et al. (1998) based on the MODFLOW-MT3D to study the groundwater susceptibility to DBCP (1,2-dibromo-3-Chlo-propane) contamination in Fresno County, California.

b) Statistical: Involves correlating actual water quality data to spatial variables and requires a large amount of site-specific data. It was developed by Jury & Roth (1992) These approaches are based on stochastic algorithms and are represented using special probability density functions for the solution of groundwater migration problems such as the Monte Carlo method (Faybishenko et al. 2013). Evans (1995) used this approach for statistical evaluation of the susceptibility of groundwater to nitrate pollution.

DRASTIC Groundwater Vulnerability Model

Among the Scoring and index approaches the DRASTIC model is one of the earliest methods for determining aquifer vulnerability and the first point count system models (PCSM) or parameter weighting and rating methods. And the most widespread PCSM method. There are two DRASTIC approaches, an ordinary DRASTIC index which can vary within the range of 23–230 or 26–260 for agricultural DRASTIC. In the case study agricultural DRASTIC model was selected as a suitable method of choice for large-scale agricultural areas study and in areas where the availability of data is low, but the general hydrogeological specifications are known according to Margane (2003); and Foster & Hirata (1988). Aller et al. (1987) created the DRASTIC model for the United States Environmental Protection Agency (EPA), to develop a framework that would allow any hydrological setting to systematically determine the potential for groundwater contamination. The DRASTIC system has two components: the classification of mapped units, called hydrogeological settings; and the implementation of a scheme for the relative ranking of hydrogeological parameters, called DRASTIC, which enables the determination of the relative potential of any hydrogeological setting to contaminate groundwater. Though the two parts of the DRASTIC system are interrelated, they are addressed in a logical progression separately. A hydrogeological setting is a composite definition of all the major geological and hydrological factors affecting and controlling the flow of groundwater into, and out of an area. It is characterized as a mappable unit with common hydrogeological characteristics and, as a result, a common vulnerability to pollution caused by pollutants introduced. The second part of the system was determined to be the seven DRASTIC parameters according to the most important mappable factors which control the groundwater pollution potential.

These variables have been arranged to form an acronym, DRASTIC for quick and easy reference, where the position of each parameter in this arrangement does not reflect its importance to other parameters. The model yields a numerical index deriving from the ranges, ratings, and weights assigned to the model parameters. Each DRASTIC parameter is subdivided into ranges or relevant forms of media types, which are ranked between 1 and 10 based on their relative effect on the potential for contamination (Aller et al. 1987). Depending on their relative significance the seven parameters are assigned weights ranging from 1 to 5, The most important factors are 5 weights; the least significant is 1.

The DRASTIC index is then determined by applying a linear combination of all variables, according to the following governing equations:

$$\text{DRASTIC Index (Pollution Potential)} = \text{DrDw} + \text{RrRw} + \text{ArAw} + \text{SrSw} + \text{TrTw} + \text{IrIw} + \text{CrCw} \tag{1}$$

where r = rating and w = weight D= Depth to water, R= Net recharge, A= Aquifer media, S= Soil media, T=Topography, I= Impact of the vadose zone media, C= Aquifer hydraulic conductivity.

$$\text{Agricultural (Pesticide) DRASTIC} = 5\text{Dw} + 4\text{Rw} + 3\text{Aw} + 5\text{Sw} + 3\text{Tw} + 4\text{Iw} + 2\text{Cw} \tag{2}$$

$$\text{Ordinary or Generic DRASTIC} = 5\text{Dw} + 4\text{Rw} + 3\text{Aw} + 2\text{Sw} + 1\text{Tw} + 5\text{Iw} + 3\text{Cw} \tag{3}$$

The only way agricultural (Pesticide) DRASTIC varies from generic DRASTIC is by allocating relative weights for the seven DRASTIC variables. All other aspects of both indexes are similar; the ranges, scores, and implementation instructions are similar. The sum of the Pesticide DRASTIC parameters weight is 26 and is also called Agricultural DRASTIC, while the sum of the Generic DRASTIC parameters weight is 23.

According to Aller et al. (1987), agricultural (Pesticide) DRASTIC is recommended to be used where agricultural activity is of concern in the study area.

Parameter ratings are derived from data on each parameter, then to be classified into ranges while each range assigned a rate according to DRASTIC ratings and ranges found in form of tables or scaling graphs of ranges and ratings (Figure 3-13) suggested by Aller et al. (1987). The significance weights are contained in the governing DRASTIC equations (equation 2, and 3), which lists weights for factors of higher applicability (Aller et al. 1987). The higher the DRASTIC rating, the higher the risk for relative contamination. The DRASTIC index can be further divided into four categories: low, moderate, high, and very high.

DRASTIC Potential Uses

DRASTIC provides a measure of the relative vulnerability of groundwater to pollution and may, therefore, be one of the many criteria used in the decision-making process but should not be the sole criterion. An example of proper use of DRASTIC would be the use of the system as a screening tool or a hydrogeological zoning map to determine whether the area is generally vulnerable to the release of contaminants on the surface. DRASTIC can be used for preventive purposes by giving priority to areas where groundwater protection is of paramount importance, particularly in arid areas.

Besides, DRASTIC can be used to classify places where special attention or protection efforts are needed. DRASTIC may be used to identify land-use practices concerning the production of contamination liability policies and the economic impact assessment of disposal costs in highly vulnerable areas. DRASTIC cannot be used in all potential applications replacing site-specific investigations or avoiding the consideration by a professional hydrogeologist of factors that may be relevant to the study area. While DRASTIC can be a very useful tool, the greater the likelihood of problems with resulting accuracy, the further the application strays from the assumptions inherent in the methodology (Aller et al. 1987).

DRASTIC can be an incredibly valuable method if the assumptions of the Methodology are completed. The four main assumptions listed by Aller et al. (1987) are:

1. The contaminant is introduced at the ground surface.
2. The contaminant is flushed into the groundwater by precipitation.
3. The contaminant has the mobility of water.
4. The area evaluated using DRASTIC is 100 acres or larger. In considering areas of 100 acres or greater, DRASTIC attempts to

determine Potential groundwater pollution from a regional perspective, rather than a site-specific focus.

The Drawback of the DRASTIC Approach

The model's results do not equate to reality by 100%. However, to simplify and provide approximate good results and explanations from the model, all the potential constraints and factors affecting were considered. It is worth mentioning the model definition by Box & Draper (1987) "all models are approximations, essentially, all models are wrong, but some are useful. However, the approximate nature of the model must always be borne in mind." Besides that, Thaddeus (2009) quoted "All models are right, most are useless" which is a more positive quote. While Breiman (2001) stated, "as data becomes more complex, the data models become more cumbersome and are losing the advantage of presenting a simple and clear picture of nature's mechanism."

Even the fact that one of the DRASTIC advantages is the large number of data layer inputs which is indeed reducing the effect of individual parameter errors or uncertainties on the result (Ouedraogo et al. 2016). Several studies discussed the reduction of the drastic parameters to achieve greater accuracy at a lower cost (Barber et al. 1993; Merchant 1994; McLay et al. 2001). And still, as in any model, there is a flaw that is the inability to interpret reality as it is, when in fact, when the model input increases, the model becomes more complex and not easy to execute and the errors increase.

Secunda et al. (1998) discussed that the DRASTIC model misses important parameters like the geological lineament density and land-use layers. However, the most interesting research in the critique of the DRASTIC model with a scientifically applicable adjustment to solve the problem was carried out by Napolitano & Fabbri (1996) the problem they discussed was the unavoidable subjectivity of the set of the seven DRASTIC parameters, the ratings, and the weights used to measure the DVIM. The key downside to the DRASTIC model is the subjectivity of the determination of the rating scale and the weighting coefficients which simply was solved by the simple statistical approach to find the effective weights by Napolitano & Fabbri (1996). Doubts have also been raised regarding the selection of specified variables and the exclusion of others.

In brief, this approach was mainly questioned on the following points:

1. Too many parameters are considered in the final DRASTIC index and important parameters of groundwater vulnerability can be influenced by other parameters that do not affect vulnerability in a specific setting (Vrba & Zaporožec 1994; Merchant 1994).
2. The choice of DRASTIC parameters is based on qualitative intuition and not on quantitative data analysis (Garrett et al. 1989).
3. Several doubts were affirmed by Rosen (1994):
   a) Several important scientifically specified parameters, such as sorption capacity, groundwater travel time, and water dilution, are not explicitly considered in the model;
   b) In contrast to aquifers in fractured media, the DRASTIC model appears to overestimate the vulnerability of porous media aquifers;
   c) It is difficult to carry out a model accuracy-test since it requires that a pollutant with the characteristics expected by the DRASTIC model (introduced into the ground surface, flooded into the groundwater by precipitation and water mobility) be deposited with a uniform concentration in the test area and over a substantial period of several years to allow the hydrogeological setting to react.

Despite these concerns, several of the advantages of the DRASTIC approach have been recognized:

1. The DRASTIC approach has a low implementation cost and can be implemented in broad regions due to the relatively few and easy-to-collect data needed (Aller et al. 1987).
2. The collection of several parameters and their interrelationship decreases the likelihood of missing certain significant parameters, limits the effect of an accidental error in the measurement of the parameter, and thus increases the statistical precision of the model (Rosen 1994).
3. The DRASTIC approach produces reasonably reliable results for large regions with a complex geological structure, despite the lack of measurements of precise parameters required by the more advanced methods (McLay et al. 2001).

DRASTIC Approach Developments

Through the reviewing of several recent studies that used the DRASTIC model to assess the susceptibility of groundwater contamination, the simpler statistical approach produced by Napolitano & Fabbri (1996) is the most referenced and most relevant, and many researchers still rely on it to reduce the subjectivity in ratings the DRASTIC parameters and improve dependability (Gogu & Dassargues 2000; Ramos-Leal & Rodríguez-Castillo 2003).

This statistical approach was first proposed by Lodwick et al. (1990) to implement a Map Removal Sensitivity Analysis (MRSA) to classify the significant and less important of the seven DRASTIC parameters so that DRASTIC users can determine if any parameters can be omitted without effectively altering the results. Following this, a single parameter sensitivity analysis (SPSA) was implemented on the same track by Napolitano & Fabbri (1996), following the same previous method but omitting one map in each sensitivity analysis, but applying the developed (SPSA) approach to the actual (or effective) weight calculation of the DRASTIC parameters used to construct a modified DVIM in this study.

In a nutshell, the next list also discusses the most important basic joints that have attempted to modify the DRASTIC model using different scientific methods over the last decades since the DRASTIC model appeared in the world of groundwater contamination risk studies.

1. Adding the land use parameter: Several earlier studies attempt of integrating land use maps as the eighth parameter in the DRASTIC model. For example, to create a vulnerability map to the Sharon region in Israel, Secunda et al. (1998) applied the DRASTIC method by adding land use parameters. Land use ratings (Lr) are arranged based on extensive land use as effluent irrigation of crops as possible sources of groundwater contamination (Table 3). The land use parameter (Lw) was assigned a weight of "5," due to its potential impact on pollutant percolation to the groundwater table. Assigned ratings and weightings for the extensive agricultural land use parameter were added to the final DRASTIC Index (DI) to produce a composite DRASTIC–Extensive-Land-Use-Index (CDI) for each cell i. The modified DRASTIC index was created by the formula $CDI_i = DI_i + Lr.Lw_i$. Parallel to this study and for the same region (Secunda et al. 1998; Melloul & Collin 1998) developed an index of aquifer

water quality and used this index to test the validation of the DRASTIC map.

Table 3. Ratings of land use categories as modified by Secunda et al. (1998)

Site-specific land usage	Ratings	Extensive land usage	Ratings	Extensive land usage	Ratings
Toxic – waste disposal	9	Cotton	10	Orchards of other fruit	6
Oil spillage	8	Built-up areas	8	Pasture or other lands unsuitable for agricultural use	5
Industries	7	Irrigated fields crop	8	Uncultivated land	5
Solid-waste disposal (regional)	6	Greenhouses/tomatoes	8	Vineyards, Olives	5
Domestic-waste disposal (local)	5	Reservoirs	7	Non-irrigated fields	4
Effluent irrigated fields	4	Citrus orchards	7	Forests, Natural areas, or reserves	1
Effluent reservoirs	3	Orchards of other fruit	6	Dune sand – open areas	1

2. Adding the land use parameter according to nitrate pollution index: Like the approaches to create the nitrate contamination index for groundwater risk assessment (GRA) introduced by Ramolino (1988), which have been explained by Canter (2019); Ramolino & Canter (1990); Canter (1987). This approach distinguishes the Nitrate Pollution Index for Groundwater based on the ranges and scores for nitrogen fertilizer (Table 4). The land-use layer (Nitrate Pollution Index (NPI)) can be created by the following formula: NPI= Ffi + Rri + Ssi + Ddi … Where, F, R, S, and D= importance weights for Nitrogen fertilizer (F), net recharge (R), Soil texture (S), and depth to groundwater (D). fi, ri, si, and di= factor ratings for the four considered factors.

Table 4. Ranges and Ratings for nitrogen fertilization by Ramolino (1988)

Range	Ratings Fi
Over-fertilized	10
Fertilized to meet crop needs	6
No fertilizer applied	1

3. Simple Land use categories ratings: Hussain (2004) modified the previous modification of integrating the land use in a simple approach to solving the limited available information about the land use and soil quality in the study area based on the groundwater quality observations, while the qualitative ratings were proposed for only three general types of land use (Table 5).

Table 5. Land use categories ratings modified in a simple approach (Hussain 2004)

Land use	Ratings
Urban	10
Rural and agriculture	8
Forest	1

4. Adding geological lineament maps: Lee et al. (1998) created a vulnerability map for Young wang County in Korea using a modified DRASTIC model by adding the "geological lineament maps" to consider the preferential migration of contamination through fractures.
5. Adding land use instead of recharge DRASTIC parameter: The National Water-Quality Assessment Program (NAWQA) to create a vulnerability map to the Eastern Snake River Plain, Idaho three of the seven DRASTIC parameters, the depth to water (D), net recharge (R) (land use) and soil media (S) were used. The land use parameter was used as a surrogate for net recharge because the studied areas having irrigated agriculture provided the largest amount of recharge (Table 6). And the final vulnerability map was correlated with nitrate concentration (NAWQA 1999).

Table 6. Land use categories instead of recharge DRASTIC parameter (NAWQA 1999)

Land Use	Ratings	Land Use	Ratings
Urban	3	Dryland agriculture	1
Irrigated agriculture	2	Forest	1
Rangeland	1		

6. Validate the DRASTIC results using the water quality: Navulur & Engel (1994); and Navulur & Engel (2003) Validated the accuracy of

the DRASTIC approach by comparing the vulnerability maps with existing groundwater quality data sampled across the Indiana state in the U.S.A.

7. Changing DRASTIC original rating: In the Turbio river valley, in Guanajuato State, the vulnerability map Ramos & Castillo (2003) modified the range of depth to water parameter by using a scale 5 times the original rating.
8. Correlation with nitrate contamination: Wei (1989) created a vulnerability map to the Fraser Valley aquifer in southwestern British Columbia (BC) and correlated between DRASTIC and AVI methods with nitrate occurrence.
9. For the case study in the next part of this chapter: Modified a simple overlaying approach to creating a land-use layer and attempts of integrating it with the modified reliable DRASTIC framework to be the parameter number eight. The DRASTIC is an overlying method to detect the intrinsic vulnerability so the DRASTIC itself was modified in this research to assess this vulnerability then the intrinsic vulnerability map was combined with the land use for the groundwater risk assessment of specific vulnerability. The particular contaminant in this approach is nitrate so the land use layer was created to reflect the potential nitrate contamination (PNC) in the study area.

Part II

Introduction

Systematic studies have been carried out by Gazal (2021b), Gazal & Eslamian (2021a), and Gazal & Eslamian (2021b) to establish an effective method for assessing the groundwater risk of renewable aquifers in the arid agricultural area by developing a multipurpose database in the GIS setting and validating the vulnerability method developed by comparing the findings with the groundwater quality characteristics of the area. Human population growth often coincides with changes in land use, including the creation of urban areas, requiring an increase in the amount of drinking water available. Groundwater is naturally abundant in the Karst and Basalt regions, where intensive agricultural areas are also situated, as in the case study area of this research.

Due to several reasons, these areas face a higher risk of groundwater contamination. Taking these aspects into consideration, a detailed groundwater risk assessment was planned in the Amman Zarqa Basin (AZB) located in the northern part of Jordan. The comprehensive groundwater risk assessment (GRA) was carried out in two steps, the first was the investigation of the contamination load and the natural groundwater's potential protectiveness. The last step was the use of the overlying modeling approaches to visualize the results of the GRA into maps. The resulted maps were evaluated by matching them with the groundwater nitrate concentration zones.

Establishing a scientific methodology that is widely and comprehensively applicable at the lowest cost is one of the most important priorities in enabling the water sector to manage groundwater resources sustainably in any water resource-poor country, intending to ensure water protection for future generations, by protecting these sources from contamination, to enhance water security and sustainability.

This comprehensive groundwater risk assessment research considered the possibility of implementing, in any region like the case study area, an integrated study to investigate the vulnerability of groundwater to contamination and to identify the contaminants in the area by the use of modern applications centered on statistical analyses, remote control, geographic information systems (GIS) and use the available data that includes water quality, hydrogeology, and climate data.

Case Study Objectives

The main objectives of this study were to assess the contamination load and groundwater natural potential protectiveness which is called a comprehensive groundwater risk assessment. And selecting the appropriate groundwater risk assessment approaches is at the core of the research. The following steps were primarily to appropriately conduct a comprehensive groundwater risk assessment research.

1. Development of a multipurpose hydrogeological database for effective storage and processing of information in the GIS environment.
2. Improving and updating the quality of input datasets through the reduction of errors and thus providing data for future groundwater management research. The data updated in this research include the

hazards, land use, soil texture, groundwater quality, and hydrological information.

3. Characterization of the geological and hydro-geological setting necessary for applying all the required groundwater contamination risk assessment analysis.
4. Assess the groundwater quality in the Study area by analyzing the long-term groundwater quality measurements of approximately 498 wells (WIS 2020) from 1970 to 2018.
5. Evaluate Jordan's and the study area's climatological parameters and generate the groundwater recharge map, by analyzing daily climatological data from 346 weather stations.
6. Investigate the state of groundwater levels and draw the water depth and groundwater flow using historical measurements from around 80 observation wells in the study area.
7. Assess the contamination load.
8. Investigate the natural protectiveness of groundwater.
9. Creating overlying modeling approaches to visualize the results of the groundwater risk assessment into maps, to be used in land use management and creating protection zones in further studies.
    a) Therefore, the first main objective of the overlying modeling part of this study was to evaluate the aquifer vulnerability in arid areas;
    b) The second main objective was to evaluate the potential risk of nitrate contamination;
    c) The third main objective in the overlaying modeling part was to integrate the intrinsic vulnerability and the potential risk of nitrate contamination;
    d) The last main objective was to validate the vulnerability and contamination risk maps by comparing the results to the observed water quality variables in the region. Aside from comparing the results to a previous vulnerability map;
    e) An additional objective was to demonstrate the combined use of DRASTIC and geographical information systems (GIS) as an effective method for groundwater pollution risk assessment and water resource management;
    f) The second additional objective was to evaluate the relative importance of the DRASTIC model parameters for assessing aquifer vulnerability in arid agricultural areas through sensitivity

analysis. And providing a comprehensive analysis of the potential application of the DRASTIC model in arid areas.

The last step in the study was the use of the overlying modeling approaches to visualize the results of the groundwater vulnerability and contamination risk on maps. The principal aim of the overlying modeling part is to present the capability of different techniques in groundwater vulnerability mapping for groundwater management against contamination in an arid agricultural area, which is in the northeastern part of Jordan. In addition to reviewing the importance of vulnerability mapping in the case of Jordan. And the resulted maps were evaluated by matching them with the groundwater nitrate concentration zones. While the model's results do not equate to reality by 100%, to simplify and provide approximate good results and explanations where the more complex the model is and considering all the potential constraints and factors affecting the model case, the more plausible the results are. According to the famous statistical book (Box & Draper 1987), "all models are approximations, essentially, all models are wrong, but some are useful. However, the approximate nature of the model must always be borne in mind." Besides that (Thaddeus 2009) quoted "All models are right, most are Useless" which is a more positive quote. While (Breiman 2001) stated, "as data becomes more complex, the data models become more cumbersome and are losing the advantage of presenting a simple and clear picture of nature's mechanism."

Problem Statement

The main problem of the present case study is tackling the deterioration of the groundwater aquifer in scarce water regions with water shortages facing climatic change effects and intensive anthropogenic activities. This requires integrated analysis and management tools, and a clear understanding of the interactions between the groundwater aquifers and the human surface activities (Dawoud & Sallam 2012). The main problem in arid and semiarid regions is the intensive water use for agriculture (Balon & Dehnad 2006) that considered the main factor causing the deterioration of the limited water resources. Arid and semi-Arid regions challenge over-exploitation-water problems that in hazardous accelerating rate in Jordan due to the refugee waves, displaced persons, and returnees from unstable surrounding countries due to prolonged conflict in the Middle East that have significantly contributed to the abnormal population increases. Besides, the rapid increase of civilization and the increasing water demands for all sectors. Continued integrated evaluations of groundwater quality and quantity are required for the sustainable development of groundwater aquifers around the world and, essentially, in arid and semi-arid areas where there are no available alternatives to water resources, while selecting appropriate groundwater risk assessment approaches considered the core of the current research.

Study Area General Description

Several factors were considered in the selection of the study area. The data availability, the importance of the study area, the extent and outcropping of the phreatic aquifers throughout the area and to be a good representative example of an arid agricultural area. For this purpose, a review of the hydrogeological situation of all the groundwater basins in Jordan was carried out to find the most important basin to be the study area. Jordan is divided into 15 surface water basins according to the topography (BGR MWI 2019) and into 12 groundwater aquifers according to the groundwater movement as outlined by Schmidt et al. (2004) and updated recently by MWI & BGR (2018). The most important and usable basin is Amman Zarqa Basin (AZB) (Figure 7). Amman Zarqa groundwater Basin (AZB), situated in the north-eastern part of Jordan, has a total area of 4104.7 km^2, while the area of the Amman Zarqa surface basin is 3588 km^2. The increasing population growth and economic activities within the study area (AZB) increase the abstraction from the main shallow groundwater aquifer to satisfy the rapidly increasing water demands which causes and continuous increase in the depth of the renewable groundwater table.

Abstraction in the Amman Zarqa Basin (Figure 8) increased gradually from 8.46 MCM in 1965 to 123.4 MCM in 1998 (MWI/USAID/ARD 2001) and it reaches 203.211% of the safe yield in Amman Zarqa Basin by a 177.81 MCM in 2019, (WIS 2020), to be at the top of the groundwater basin ranking table of groundwater abstraction, with having the greatest portion of Jordan's production wells, see the table of abstraction from the groundwater basin (Table 7).

AZB is the most important part of and groundwater basin in Jordan as it is a concentrated human activity area, where 85% of the industries (Al-Mashaqbeh et al. 2014; Daniel et al. 2013) and more than half of the country's population is in this basin which is only around 4% of the total country area

(DOS 2020). The potential for water quality and quantity degradation of the phreatic Aquifer in the study area is high, due to several factors, especially the increase in agricultural water demands in the AZB. As agriculture is the largest water consumer in the basin according to the National Water Master Plan (NWMP) (MWI 2004; MWI 2016a), with continuous population growth and improving living standards, that results in rapidly increasing per capita water demands. The boundary of Amman Zarqa Basin (AZB) was considered to sketch the boundary of the study area, which was carried out in the eastern part of this most important and populated groundwater basin in Jordan, which is characterized by a rapid agricultural expansion that is raising the threat of groundwater contamination according to several comprehensive hydro-geological studies like (MWI/USAID/ARD 2001), (MWI & JWA 2010), (Daniel J. et al. 2013), and (MWI 2013). The study area covers a surface area of 3329.9 km^2 (more than 81% of the (AZB) area, and 3.7% of Jordan's land area). According to the administrative divisions' maps recovered from the Royal Jordanian Geographic Centre (RJGC 2020) in a shapefile format, the study area is governed by four governorates, Mafraq, Zarqa, Amman, and Jerash, accounting for 57%, 28%, 15% and 0.4% of the study area, respectively. And divided into districts and sub-districts where the study area is administratively subordinate to (11) districts (Figure 9).

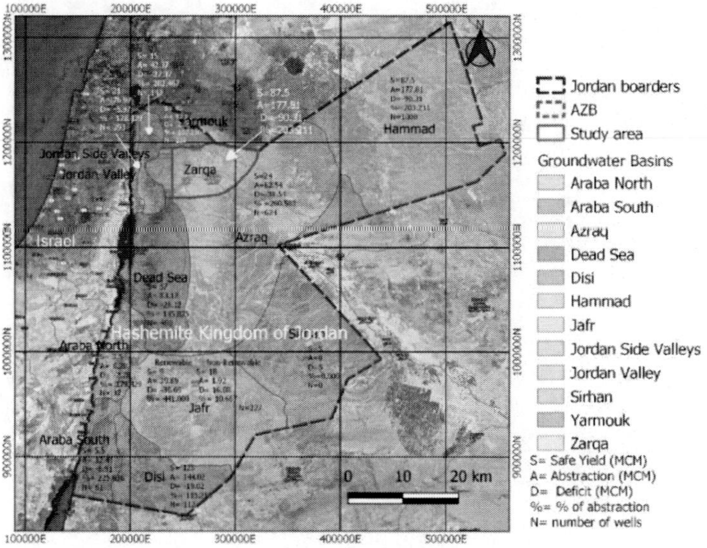

Figure 7. Groundwater basin in Jordan, and the annual abstraction. (Own processing) retrieving the raw data from WIS (2020).

Table 7. Groundwater abstraction (2019) and the safe yield values according to the latest updated estimations (BGR & MWI 2019), and the number of wells in each basin according to WIS (2020)

*	Groundwater Basin		Safe Yield (MCM)	Abstraction (MCM)	Deficit (MCM)	% of abstraction	Number of wells
1	Amman-Zarqa (AZB)		87.5	177.81	-90.31	203.211	1009
2	Azraq		24	62.54	-38.54	260.583	624
3	Dead Sea		57	83.12	-26.12	145.825	480
4	Disi Non-Renewable**		125	144.02	-19.02	115.216	112
5	Hammad		8	1.34	6.66	16.750	9
6	Jafer	Renewable	9	39.69	-30.69	441.000	227
		Non-Renewable**	18	1.92	16.08	10.667	
7	Jordan Valley		21	26.97	-5.97	128.429	393
8	Jordan Side Valley		15	42.37	-27.37	282.467	142
9	Sirhan		5	0	5	0.000	0
10	Araba North		3.5	6.28	-2.78	179.429	37
11	Araba South		5.5	12.41	-6.91	225.636	61
12	Yarmouk		40	42.9	-2.9	107.250	228
	Total		418.5	641.37	-222.87	153.2544803	3322

*Jordan consists of 15 surface water basins and 12 groundwater basins. **The safe yield abstraction from non-renewable groundwater for 50 years is about 143 MCM annually according to the Ministry of Water and Irrigation studies (MWI 2016).

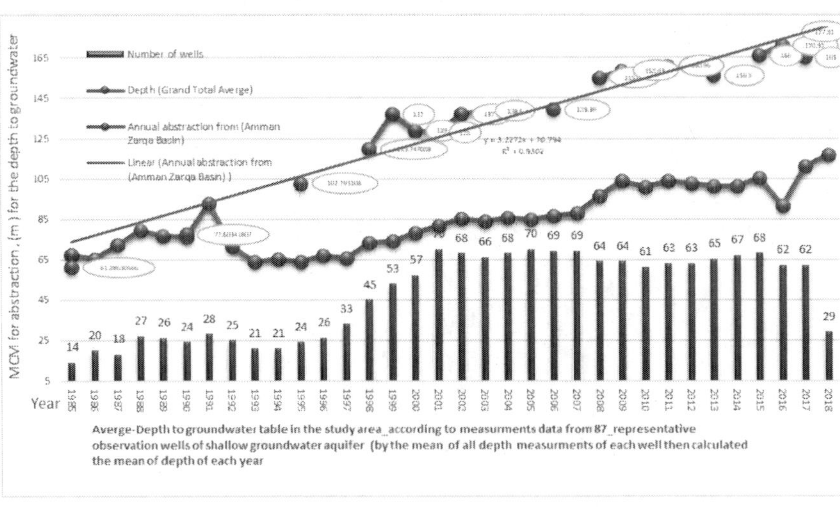

Figure 8. Depth to groundwater table in the study area, and the annual groundwater abstractions, (own processing).

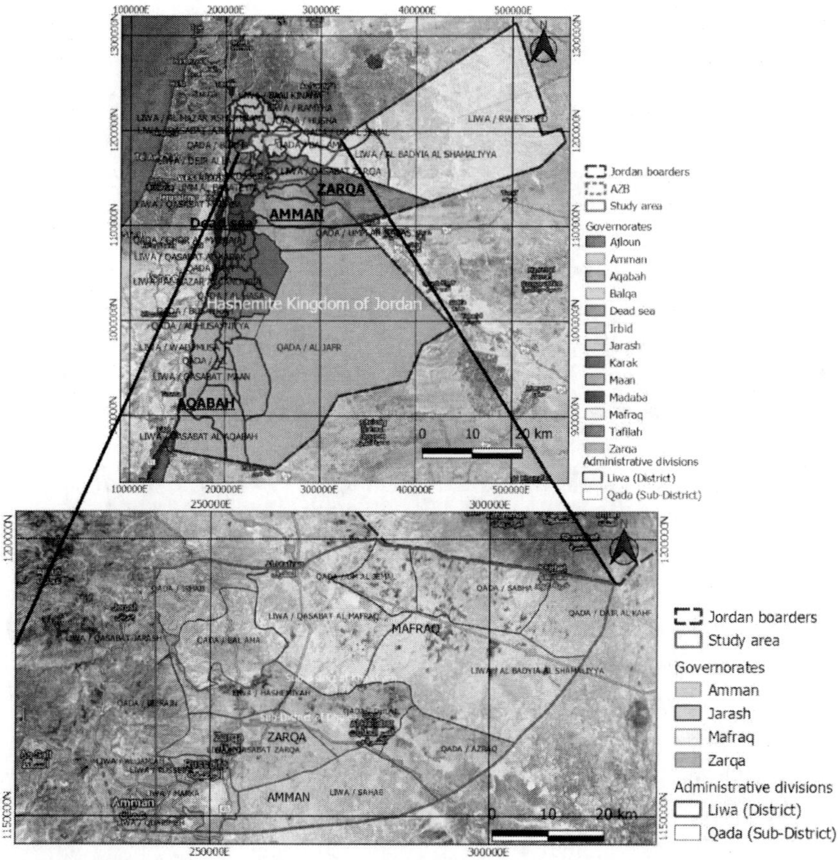

Figure 9. Study area location and the administrative divisions according to (RJGC 2020).

Methodology

The comprehensive assessment of groundwater risk in the case study is defined in the following scheme (Figure 10). The groundwater risk assessment (GRA) approach in this study consists of three steps, the contamination load investigations, then the comprehensive hydrogeological review. The last step is to visualize the groundwater vulnerability and the nitrate contamination risk in maps utilizing the overlying modeling techniques. Because vulnerability modeling is the focus of this chapter, a brief overview of the first and second steps' outcomes will be provided in the next paragraphs, while the third step will be completely examined in the following sections. The methodologies applied in the investigations of both GRA components (the intrinsic vulnerability and the contamination load), can be summarized by the following:

1. Investigation of the contamination load in the study area.

The major human activities on the surface of the study area are farming activities according to several official and scientific studies, but in this study, a review of the study area contamination sources have been carried out for more precise findings. In this study, the following analyses were done to investigate the water quality in the study area:

- A summary table was drawn up and discussed for the analysis of the main water parameters;
- Correlation analyses of the main water parameters;
- A Piper diagram of the major cations and anions was used to sketch the main cation and anion concentration of the studied wells using the Plots Piper diagram model;
- Three parameters were selected nitrate, sulfate, and the total dissolved solids to prepare spatial continuous concentration maps, to evaluate the water quality in the study area, and to compare it with the created vulnerability maps.

2. Investigation of the natural protectiveness potential of groundwater in the study area.

In this part, the aim is to provide an overview of all available hydrogeological and climatic parameters that contribute to the GRA. This is to construct a better understanding in the investigation of the natural potential protectiveness of the groundwater. In this study, the following analyses were done to investigate the natural protectiveness in the study area:

- Climate, Geological, and Hydrogeological Systems Description of the case study area;
- Structural geology and the Hydrogeology of the study area;
- Groundwater Recharge in the study area;

- Related physical features of the landscape obtained from remote sensing (RS) Technique for Groundwater Condition Assessment;
    a. Geological Lineaments and Structural setting;
    b. Surface water drainage density;
- Hydraulic conductivity of the aquifer system in the study area.

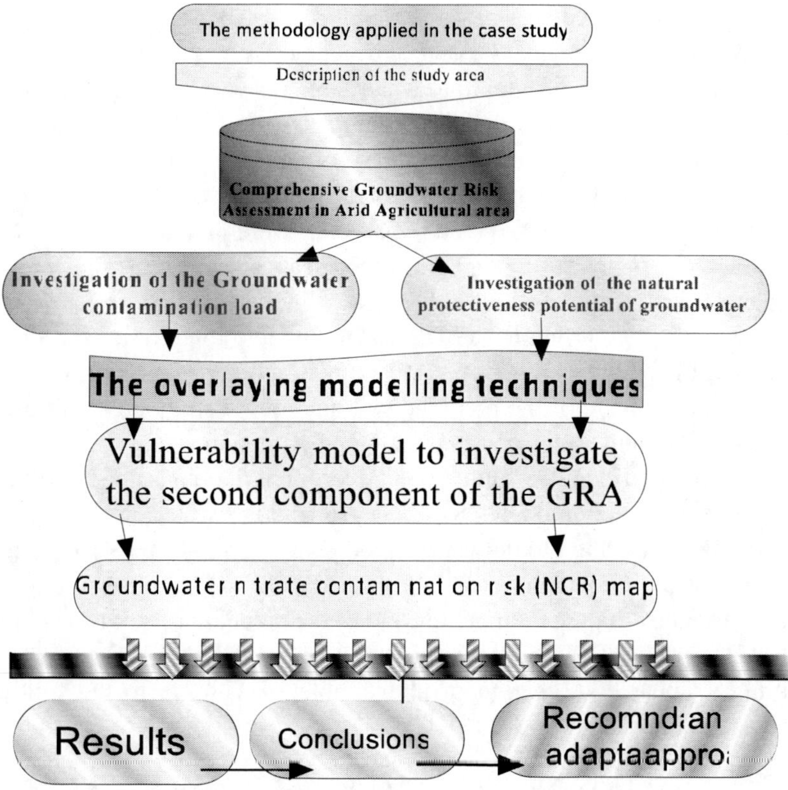

Figure 10. The general methodological scheme of the study.

The Overlaying Modeling Techniques

To simulate the previous investigations of both GRA components (the intrinsic vulnerability and the contamination load) overly modeling techniques were used as explained (Figure 11). Three overlays modeling steps were performed to create the groundwater intrinsic vulnerability map, potential nitrate contamination (PNC) map, and the groundwater nitrate contamination risk (NCR) map.

Investigate the Second Component of the GRA by Modified Intrinsic Vulnerability Model

Figure 12 summarized the two DRASTIC approaches applied in this research. The agricultural and Ordinary (generic) DRASTIC approaches were used in this study to develop the groundwater intrinsic vulnerability map in addition to the former intrinsic vulnerability map created by COP-model as set out in (Vías et al. 2006). (Aller et al. 1987) created the DRASTIC model for the United States Environmental Protection Agency (EPA). The DRASTIC seven parameters thematic maps, depth to water table (D), net recharge (R), aquifer media (A), soil media (S), topography (slope%) (T), the effect of the vadose zone (I), and hydraulic conductivity(C) were prepared relay on the hydrogeological, soil, groundwater recharge, depth to the water table and the created slop maps. Where the (A), (I), and (C) were prepared by referring to the geological maps utilizing the geological attribute of the prepared hydrogeological map of the study area which was used to encode the geological units according to the DRASTIC model rating system for aquifer media (A), impact to vadose zone (I), and hydraulic conductivity (C). While the soil units attribute of the prepared soil map of the study area was used to encode the soil units according to the DRASTIC model rating system for soil (S). The prepared depth to the water table, recharge, and slope% thematic layers were classified into DRASTIC (D), (R), and (T) rates, according to the ranging of depth to the groundwater table, recharge, and slope % (Figure 13).

The model yields a numerical index deriving from the ranges, ratings, and weights assigned to the seven parameters of the model. According to (Aller et al. 1987), the DRASTIC index vulnerability map (DIVM) is determined by applying a linear combination of all variables, according to the following governing equations:

$$DRASTIC\ Index\ (Pollution\ Potential) = \mathrm{DrDw + RrRw + ArAw + SrSw + TrTw + IrIw + CrCw} \tag{1}$$

where r = rating and w = weight D= Depth to water, R= Net recharge, A= Aquifer media, S= Soil media, T=Topography, I= Impact of the vadose zone media, C= Aquifer hydraulic conductivity

$$\mathrm{Agricultural}\ (P\mathrm{esticide})\ \mathrm{DRASTIC} = \mathrm{5Dw + 4Rw + 3Aw + 5Sw + 3Tw + 4Iw + 2Cw} \tag{2}$$

$$Ordinary\ or\ (Generic)\ \text{DRASTIC} = 5Dw + 4Rw + 3Aw + 2Sw + 1Tw + 5Iw + 3Cw \tag{3}$$

Agricultural (Pesticide) DRASTIC is a particular case of the DRASTIC Index.

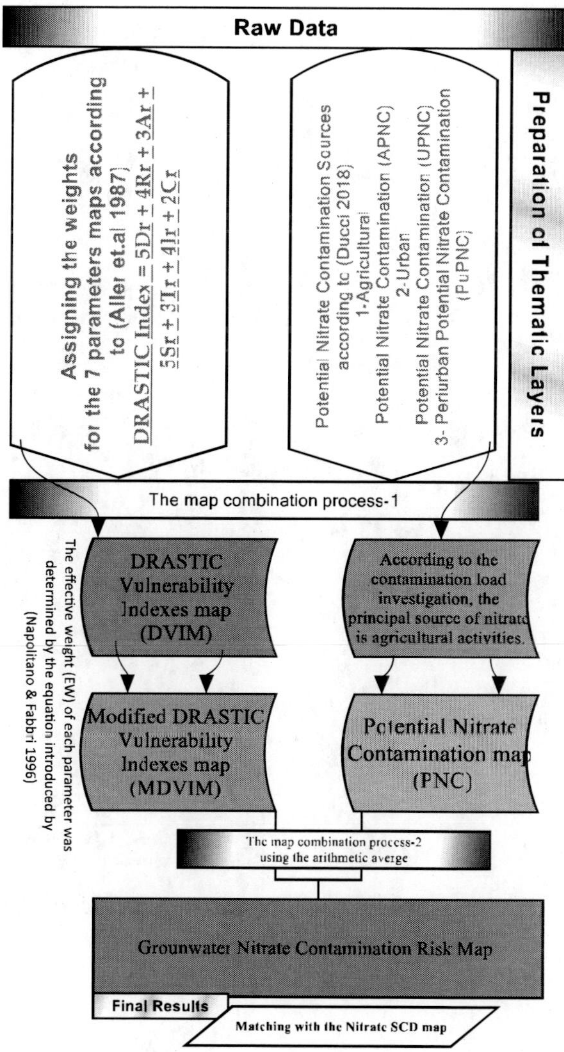

Figure 11. Flow chart simplifies the overlaying modeling techniques used in the case study.

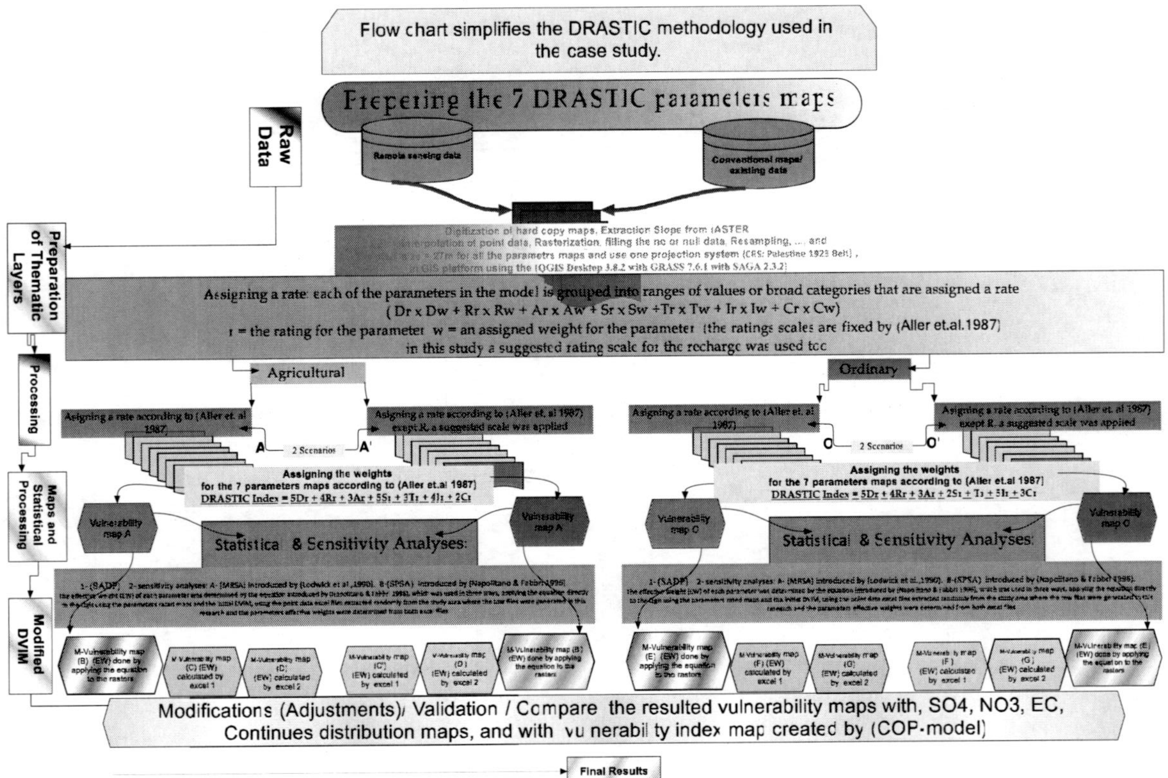

Figure 12. The schematic flow chart describes the DRASTIC-model approaches used for this study to create an appropriate intrinsic vulnerability map.

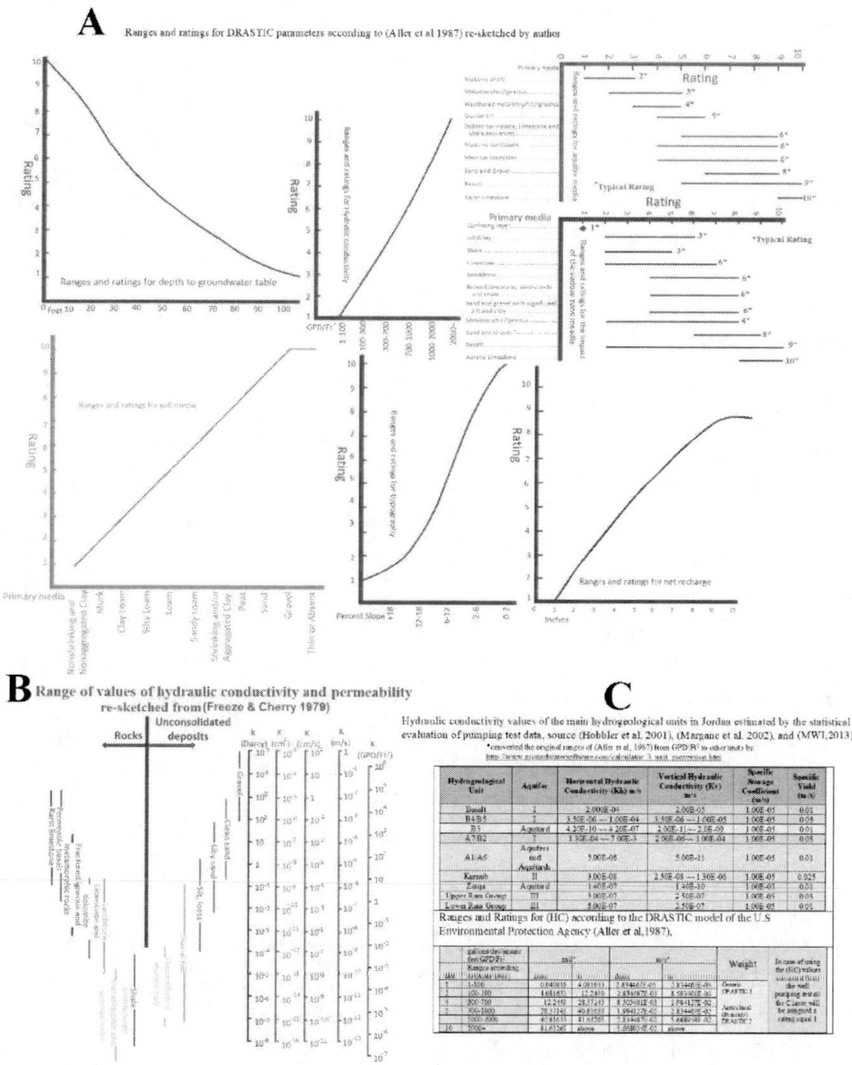

Figure 13. A: Ranges and ratings for DRASTIC parameters according to (Aller et al. 1987). B: Range of hydraulic conductivity and permeability values (Freeze & Cherry 1979). C: Wrong Hydraulic conductivity values estimated from wells pumping tests.

Agricultural (Pesticide) DRASTIC is recommended to be used where agricultural activity is of concern in the study area (Aller et al. 1987) so the last decision was to use the result obtained by performing this approach. The preparation of the DVIM and the statistical and sensitivity analyses to modify it was processed in the Geographical Information System (GIS) and was performed in the free available software used in this study (QGIS Desktop 3.8.2 with GRASS 7.6.1 with SAGA 2.3.2).

In this study, the two DRASTIC approaches were performed based on equations 2 and 3. There were two scenarios in each of the two approaches as mentioned in (Figure 12). The last decision, however, was to follow the Agricultural Approach. The next steps clarify the development of vulnerabilities maps through the implementation of the Agricultural Approach, although the same steps have been implemented in the Ordinary DRASTIC approach.

Development of the DRASTIC Seven Thematic Layers

Several data types and sources were used to create thematic layers of the seven DRASTIC parameters. All the DRASTIC parameters were subdivided into ranges according to (Aller et al. 1987) and have been assigned a relative weight according to equation 2. Each DRASTIC parameter was processed in the Geographical Information Systems (GIS) and was performed in the (QGIS Desktop 3.8.2 with GRASS 7.6.1 with SAGA 2.3.2) software and prepared as thematic layers. The DRASTIC approach in this study was performed by two scenarios, A and A´, the first one with the original ranges and ratings the second one only the ratings of recharge were suggested to be changed as discussed in the preparation of (R) parameter below. The results from the DRASTIC vulnerability index map (DVIM) from the two scenarios (A and A´) were followed by complete statistical and sensitivity analyses and were modified by the calculated effective weight of the DRASTIC parameters as described below.

Depth of groundwater (D), according to the DRASTIC model, (D) the distance from the ground surface to the groundwater table in an unconfined aquifer is one of the important factors in assessing the area's most vulnerable to contamination where, with a higher groundwater depth, there is a lower risk of contamination. A deeper groundwater table implies longer travel times, as well as the longer the passage period for the water that flows to the groundwater table from the surface, the greater the likelihood that the water will get rid of part of the pollutants that it may carry from the surface, therefore areas with a large depth of water are less prone to contamination. The depth to groundwater map was prepared after creating the static water level (SWL) map (Figure 14) by ordinary kriging (OK) interpolation of the observation well's SWL average measurements.

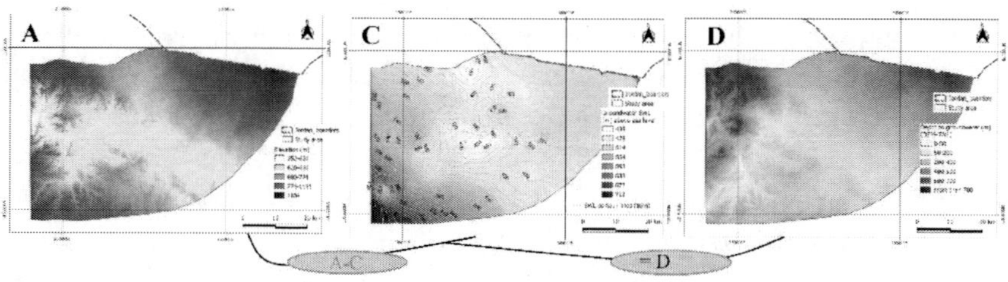

Figure 14. Creating the groundwater depth map.

Then Subtract the SWL map from the DEM thematic layer to obtain the depth of the groundwater table map. Then the depth to the groundwater table map was classified into the DRASTIC (D) rates, according to the range of depth to the groundwater table, the minimum rate assigned by DRASTIC is one reflecting a low groundwater risk while the maximum rate is ten reflecting a high groundwater risk of contamination. The depth to groundwater of more than 99% of the study area belongs to the lowest risk DRASTIC (D) range (D >39) therefore almost all the study area was rated by one, so the dominant scoring index of D is 5.

Net Recharge (R), groundwater recharge (GWR) is the estimated annual amount of water that infiltrates from the ground surface into the aquifer. The natural GWR is considered one of the most important factors in groundwater risk assessment as a significant medium for transporting the contaminants, whereby pollutants are transferred to the groundwater, and the higher the recharge percentage, the greater the groundwater risk. The GWR map of the study area was created from Jordan's average GWR map, which was prepared by (Gazal 2020a) according to the Jordan recharge approaches of (Hobbler et al. 2001), the percolation rate was determined based on rainfall-recharge percentages equations which were previously adopted and used by (MWI 2019). These equations were applied depending on outcropping formations in Jordan, in the study area, for example, the adopted average recharge percentage of rainfall for the outcropping geological formations according to (Hobbler et al. 2001), Basalt, B4, B3, A7/B2, A1/6, Kurnub, are (15%, 7.5%, 0, 15%, 7.5%,7.5% respectively). According to the study area thematic GWR layer, the mean GWR is 14 mm/year, the maximum 42.6mm/year, the minimum is zero in the south part as the outcropping formation there is the B3 which is aquitard. According to the average precipitation rate of the study area which is 134mm/year, the calculated GWR to precipitation in the study area is 10.4% which is a good percentage. But the precipitation rate is very small in the study area, so according to the DRASTIC ranges and ratings, when the

GWR ranges between zero to.05m/year the assigned rate is one (as low-risk potential from the GWR point of view). Therefore, the area was entirely rated 1, therefore the second scenario was suggested by changing the recharge ranges and ratings (R'), by multiplying the ranges by 100 (Table 9). These suggested ranges and ratings of (R') are intended to simulate the potential increase in recharge by irrigation return flow (IRF) in intensive agricultural areas. However, the agricultural activities are not only increasing the GWR through (IRF) but the processes of ploughing that disintegrate the soil layer which increases permeability and enhancing the recharge.

Aquifer Media (A) refers to the (consolidated or unconsolidated materials) characteristics of the saturated zone that control the process of attenuation of the pollutant. The ranges and ratings of the DRASTIC are assigned according to the water travel time in the aquifer, the longer the travel time will result in more attenuation of the pollutant and the rate is in reverse relation to this travel time. Each Aquifer media in the ranges and rating scale of DRASTIC is listed in the order of increasing pollution potential, the aquifer media with the highest pollution potential is assigned the highest scores of the rate as suggested for karstic and basalt aquifer 10, while in this study the suggested typical DRASTIC rates for basalt and karstic were used 9,10 respectively. And so, the other aquifers media in the study area were assigned rates, according to the geological description of each unit then return to the scale of ranges and ratings to assign a representative scoring rate as in (Table 4-11). for example, the scoring rates of sandstone media can be between 6-9 according to the (Figure 13) while the choice is governed by the specific characteristics and properties of this media, the sandstone aquifer (K) in the study area was assigned a scoring rate of 8 which is the typically suggested rating score for sand and gravel aquifer media by (Aller et al. 1987). Both Aquifer (A) and hydraulic conductivity (C) thematic layers were prepared by referring to the geological formation in the study area, and the geological attribute of the prepared hydrogeological map was used to encode the geological units according to the DRASTIC-model rating system for aquifer media and hydraulic conductivity.

Soil (S), soil media refers to the vadose zone's uppermost part, marked by substantial biological activity (Aller et al. 1987). The prepared study area soil map (Figure 3-8) was used to create the thematic layer of a DRASTIC rated soil map, which include an attribute of the texture class of each soil unit according to the soil textural classification triangular chart (USDA 2014), therefore each soil unit was rated according to DRASTIC soil ranges and ratings. There are 5 soil classes in the study area the dominant class is the Silty clay loam which was assigned a scoring rate of 3 according to DRASTIC ranges and ratings, so by multiplying with the soil weight then the dominant scoring index of the DRASTIC S (wSr) is 15 (Table 4-11).

Topography (T) parameter refers to the inclination degree of the ground surface (terrain slope), land surface slope variability as an input in the DRASTIC model expresses the flow rate at the surface in the approach of determining the most susceptible areas to pollution. The slope parameter determines how much the water stays on the surface. Whereas water stagnates in low slope areas, increases infiltration, and there is a greater potential for contaminant migration. While the sloping areas with a higher inclination are considered less susceptible to pollution because of the periods of water settling on the surface not enough to high rate of water infiltration to the water table, including the contaminants it contains potential presence due to agricultural activities, for example. A slope degree map was created using data from the digital elevation model (DEM) derived from the high-resolution Shuttle Radar Topography Mission (SRTM) (NASA 2019) (Figure 14). according to the slop degree thematic layer, only 0.016% of the study area has a slope degree above 18 degrees and the degree of slope exhibited by the study area varies from 0 to 32.43 degrees and the standard deviation of the slope degree map is 4.249 and the coefficient of variation is 0.965 (before classified it into DRASTIC scoring ratings according to ranges of slope degree). Therefore, the study area is considered a gentle slope area with an average slope degree of 4.4. Then topography (T) parameters are prepared by classifying the slope degree map into five classes according to ranges and ratings (Aller et al. 1987). Areas with low slopes (<2%) were typically assigned a high rating score (Tr=10) indicating their high effect on the aquifer vulnerability, thus the scoring index of the highest rating was calculated by multiplying the rating score (Tr) by the (wT) then the scoring index is 30 the highest T DRASTIC index (Table 4-11).

Impact of vadose zone (I) refers to the hydrogeological characteristics of the vadose zone materials, which are the discontinuously saturated or unsaturated layers above the water table of the renewable aquifers. These characteristics determine the process of attenuation of the pollutant within the distance from the soil layer down to the water table of the saturated aquifers. (Aller et al. 1987) classified the vadose medium into an order of increasing groundwater pollution potential. The (I) parameter map was prepared from the 14 structural geological maps retrieved from (MEMR 2017), compiled, processed and the geological attribute was used to encode the vadose zone formations with rating scores according to (Aller et al. 1987). In the study area, the highest DRASTIC (I) risk rating score was assigned to the Wadi As sir limestone (A7 formation), as this formation is characterized by a highly fractured and karstification, thus the highest contamination risk index of I (wIr) is 40 as the weight of the I is 4 (Table 4-11).

Hydraulic conductivity (C), according to (Aller et al. 1987), HC refers to the ability of aquifer materials to transmit water, which in turn regulates the rate at which groundwater flows under a given hydraulic gradient. The rate at which groundwater flows also influences the rate at which a contaminant is moving away from the point at which it reaches the aquifer. HC was inferred from the range of values of HC and permeability (Freeze & Cherry 1979), taking into consideration the geological characteristics of the aquifers discussed before, which assure that the dominant aquifer media in the study area are the karstic highly-fractured limestone of B2/A7 formations and the basaltic formations both are highly permeable so by referring to the ranges of HC both A7/B2 and Basalt are considered according to DRASTIC ranges and ratings to be assigned a scoring rate of 10 for A7/B2 and basalt, as they belong to the highest risk DRASTIC (C) range (>2000m/s), so the dominant scoring index for C (wCr) in the study area is 20 as the Cw is 2. In the original (Aller et al. 1987) ranges and ratings of parameter C, the HC ranges were plotted using GPD/Ft^2 unit (Table 3-4), but the unit of the estimated HC values from the pumping well test in Jordan is (m/s). So, the HC ranges of (Aller et al. 1987) have been converted to (m/s), as listed in (Table 3-4). While according to (Gazal 2020a) and (Hobbler et al. 2001) the HC estimations from the pumping well test in Jordan is not valid for scientific purposes. Professor Hobbler's thorough explanation stated, they are impractical or constitute incorrect numbers that do not give real values to the hydraulic conductivity (HC) of the aquifers in Jordan and it was confirmed that the Jordanian water authority (WAJ) is primarily interested in well performance tests to assess the yield and efficiency not the hydraulic parameters of the aquifer. As a result, the range of hydraulic conductivity and permeability values created by (Freeze & Cherry 1979) (Figure 13) and adopted by numerous hydrogeological studies such as (Aller et al. 1987) has been used in the case study of this research, and it is suggested that this range be used where the HC values do not exist or there is doubt as to their accuracy.

DRASTIC Statistical and Sensitivity Analyses

Using many data layer inputs is one of the advantages of the DRASTIC model, such as increasing the number of data layers, reducing the effect of individual parameter errors, or uncertainties on the result (Ouedraogo, Defourny & Vanclooster 2016). Nonetheless, many studies, such as (Barber et al. 1993),

(Merchant 1994), and (McLay et al. 2001) discussed approaches to reduce the number of DRASTIC input parameters to achieve greater accuracy at a lower cost. (Rosen 1994) Statistically studied the interrelations between the seven DRASTIC parameters in several sites in Sweden and found they are quiet and representative to assess the vulnerability. However, others studied like (Secunda et al. 1998) assumed constancy of the missing other parameters as well as several recent studies include more parameters in the DRASTIC approaches such as the lineament density and land-use layers. Moreover, (Napolitano & Fabbri 1996) criticized the unavoidable subjectivity of the collection of the seven DRASTIC parameters, the ratings, and the weights used to measure the DVIM. Generally, the approaches to groundwater vulnerability assessment need validation to minimize subjectivity in ratings and range the parameters of input and improve dependability (Gogu & Dassargues 2000), (Ramos-Leal & Rodríguez-Castillo 2003).

Therefore, modifications attempts were carried out in this study to accurately determine groundwater susceptibility in the study area. Statistical and sensitivity analysis was carried out to understand the influence of the ratings and weights assigned to each of the DRASTIC parameters and calculated the effective weight of each DRASTIC parameter.

1. Statistical analyses of DRASTIC rated parameters (SADP), a compare investigated study accomplished utilizing the statistical summary of each DRASTIC rated parameter thematic layer to study the variability, in addition to a correlation analyses to examine the independence of the DRASTIC rated parameters, where the independence of the DRASTIC parameters reduces the possibility of error, while generally, the DRASTIC parameters are naturally closely related parameters according to (Rosen 1994). While the low variability of the DRASTIC rate parameter indicates a smaller contribution to the variation of the vulnerability index across the study area (Babiker et al. 2005).
2. Sensitivity analyses: Analysis of sensitivity offers useful information about the effect of rating values and weights applied to each parameter and helps to determine the importance of subjective parameters. Two sensitivity analyses were carried out in this study; the map removal sensitivity analyses (MRSA) introduced by (Lodwick et al. 1990) and the Single Parameter Sensitivity Analysis (SPSA) introduced by (Napolitano & Fabbri 1996) in addition to applying the developed (SPSA) approach (Napolitano & Fabbri 1996)

in the real or the effective weight calculation of the DRASTIC parameters which were used to create a modified DVIM.

a) *Map removal sensitivity analyses (MRSA),* introduced by Lodwick et al. (1990), the MRSA identifies the sensitivity of the vulnerability map towards removing one or more maps from the creation of the DVIM analysis, which is calculated by the following equation:

$$S = \left(\frac{\left| \frac{V}{N} - \frac{V'}{n} \right|}{V} \right) * 100 \tag{4}$$

where, the sensitivity associated with removing one or more maps (S): is the sensitivity measure expressed in terms of variation index*, (V) is the unperturbed DVIM (the initial one calculated by the DRASTIC govern equation 2), V´ is perturbed vulnerability indices, N and n, is the number of data layers used to calculate the V and v. While the $variation\ index = \frac{V-V'}{V}$. In this study 2 initial vulnerability maps were created according to the two scenarios of implementing the DRASTIC, so the test will be performed twice one with the DVIM by the first scenario (map A) the second by the DVIM created by the second scenario (map A′) (Figure 3-12).

b) *Single-parameter sensitivity analysis (SPSA),* performed to modify the weight of the DRASTIC parameters (Napolitano & Fabbri 1996), indicated that SPSA provides valuable weighting impact information allocated to each parameter and enables analysts to determine the value of subjectivity. In the discussion through the results of this study, these analyses were used to determine the effect of each DRASTIC parameter on the resulted DRASTIC vulnerability indexes map (DVIM) to evaluate whether it was really necessary to use all the seven DRASTIC parameters, the higher influence of a parameter compared by another parameter, indicate a higher real weight of this parameters. The calculated variation index of the perturbed

DVIM is the measure of the real influence of the removed parameter on the final DVIM.

In this analysis that was initiated for the first time by (Napolitano & Fabbri 1996) the equation (Lodwick et al. 1990) number 4 is used but as the number of the removed parameters is fixed by only one so the equation can be retyped the as:

$$S = \left(\frac{\left| \frac{V}{7} - \frac{V'}{6} \right|}{V} \right) * 100$$

which by the raster calculator can be written as:

S=(absolute((V/7)-((V-(X*Xw))/6))/V

(5)

where X is the removed rated parameters, Xw the parameter theoretical weight.

Each DRASTIC parameter contributes to the DVIM with an effective weight (W) which can be calculated in % for each parameter:

$$Effictive\ (real)\ W = \left(\frac{PrPw}{V}\right) * 100$$

(6)

Pioneered by Napolitano & Fabbri (1996), where V: DVIM is calculated by the original parameter weights, Pr, Pw is the ratings and the weights respectively of the parameter P.

DRASTIC Validation

To examine the vulnerability maps created by the DRASTIC model long-term groundwater nitrate, sulfate, and total dissolved solid concentration analyses were retrieved (WIS 2020). These parameters were selected considering agricultural activities as the main contamination sources in the study area, while the common chemical fertilizers used by farmers are ammonium sulfate fertilizers (Gazal 2020a; Gazal 2015). The real maximum and the average concentration measurements of Nitrate sulfate and EC (TDS in mg/L = 0.54

EC in μS/cm, 0.54 is the conversion factor used in Jordan by the MWI (JISM 2008)) were interpolated by Ordinary Kriging (OK) method to create six spatial continuous data (SCD) maps.

Lastly, the recent Jordan's intrinsic vulnerability map by COP-model (BGR & MWI 2018) was used in this study to compare with the created DVIM. COP-model as set out in Vías et al. (2006) uses three parameters, the C: Concentration of Flow (slope and surface features) the O: Overlying Layers (soil and geological formations), and P: Precipitation (quantity and intensity), which give the precipitation factor third of the weight in determining the groundwater vulnerability.

Groundwater Nitrate Contamination Risk (NCR) Map

The last step in the overlying modeling application in this study is creating the groundwater nitrate contamination risk (NCR) map as described in (Figure 3-11). A groundwater risk map generally, is a combination of existing hazardous substances and the aquifer vulnerability. But due to the difficulties of recognizing all possible sources of contaminants and the presence of a real contamination problem, the susceptibility is more directed towards a particular contaminant. While, due to the results of the contamination loads investigation, Nitrate was selected as a particular contaminant in the study area.

The first component of the NCR is the potential nitrate contamination (PNC) was created by overlapping three land-use weighted thematic layers responding to the nitrate contamination factors. According to (Ducci 2018) each layer was classified into five classes (Table 4-24), described below:

1. The agricultural potential nitrate contamination (APNC): Mainly related to the fertilizers load which is high in the areas with high agricultural intensity. While the agricultural intensity map (AIM) was created by compiling a set of AIMs recovered from (MOA 2020) in hard copy format, reflecting the AIM of the year 2019 was considered to create the APNC layer. Five APNC classes were assigned to the study area, whereas the areas with high agricultural intensity were assigned a high APNC class, while the areas with very low agricultural intensity were assigned a very low APNC class.
2. The urban potential nitrate contamination (UPNC): It is associated with the likelihood of leakage of the sewage network and is expressed by the population density which is low to very low in the study area. This layer was prepared to utilize the administrative divisions' maps

recovered from (RJGC 2020) and population data from (DOS 2020). The population density in Jordan as a whole reached 121 inhabitants/km^2 which is assigned a moderate UPNC according to the population density classification used in this study. The population density in the studied area (the eastern part of AZB) is low to very low and moderate to high in the excluded part of AZB.

3. The Peri-urban Potential Nitrate Contamination (PuPNC): Linked to the percentages of the sewer system coverage (SSC), where the areas with low SSC were assigned a high PuPNC class considering the existing illegal sewage connections coexist with on-site sewage disposal like septic tanks, while the areas with high SSC were assigned a low PuPNC class. This layer was prepared by combining the administrative divisions' maps recovered from (RJGC 2020) and the SSC percentage for each administration unit retrieved from (WIS 2020).

In the index overlay combination, considering all the above three classified land use maps of equal weight the resulting values are the arithmetic average of the input values, starting from 1 (very low) to 5 (very high) PNC classes. The final step to create the NCR map was by overlapping the PNC map and the modified DVIM map according to the equation $NCR = \frac{PNC+DVIM}{2}$.

The NCR map was matched with the nitrate SCD map to evaluate the above-adopted methodology.

Results of the Case Study

Investigation of the Groundwater Contamination Load

In the selected study area through the field visits, the hazard files, water quality analyses, classification of the production wells into purposes of uses, shows the highest portion of contamination is related to agricultural uses so it turns out that the most important potential source of pollution drives from agricultural activities.

The Study Area is Agricultural Activity

The study area agricultural activity intensity map indicated that there is intensive irrigated agricultural activity in the northeastern part of the study area (Table 8). While the study area's original natural vegetation is scattered with short plants and it is within the range proposed for rain-fed cereals crops, not intensive crops according to land suitability maps prepared by MOE & UNDP (2015). The total number of productive groundwater wells in Jordan in 2019 reached 3321 wells, 1009 of which are in the AZB and 900 of these wells in the study area, and thus more than 28% of Jordan's wells are in the study area. Most of these wells are for agricultural purposes with a percentage reaching 69% of the total number of production wells in the study area and 23% for drinking water purposes. It demonstrates the agricultural nature of human activities in the study area rather than industrial activities. To conclude in the study area, agricultural activities are the key sources of contamination by both non-point sources or by point sources such as FMB and other sources of contamination as hazard locations exist in the study area, for example, fertilizer factories and poultry farms.

Table 8. The agriculture intensity in the study area, the original map source (MOA 2020)

Agricultural activities	% of the study area
Agric. > 75%*	8.354084
Agric. 50-75%	5.923244
Agric. 25-50%	27.84274
Agric. 10-25%	3.878595
Agric. < 10%	23.63339
No agric., dom. pasture	28.59867
No agric., other land uses	1.769272

*The intensive agricultural activities cover 75% of the agricultural areas in these areas.

Analyses of Groundwater Quality Data in the Study Area

By analyzing the two water quality files it was found that in the first file for the period from 1970 to 2005 there were about 6459 nitrate measurements and about 2607 of which were above 50 mg/l, and about 7427 TDS measurements 1901 of which were above 1000 mg/l (1850 μS/cm), and 6364 sulfate measurements, 213 of which were above 500 mg/l. While the second data file from 2006-2018 there was 976 nitrate measurement 268 of which were above 50 mg/l, about 35 TDS measurements 6 of which were above 1000 mg/l, and there were 873 sulfate measurements 76 of which were above 500 mg/l. In the studied wells the long-term Sulfate concentration of less than 500 mg/l is possibly derived naturally from gypsum dissolution while the high concentration may be derived from the use of the fertilizer.

Potassium ions may have come from irrigation return flow (IRF) as the concentration varied from 0.1 to 257 mg/l with an average concentration equal to 8.5 while the commercial chemical fertilizers used in AZB have a high concentration of potassium. The nitrate concentration varied from 0.2 to 376 mg/l with an average of 46 mg/l and the phosphate ranges from 0.01 to 0.27 mg/l, the high nitrate and phosphate concentration in the groundwater of the study area is mainly related to the use of fertilizers in agricultural areas. The temporal changes in water quality by reviewing several studies show an increasing pattern of the concentration of nitrate and salinity.

Piper Diagram

According to the Piper-trilinear model, the dominant hydro-chemical facies in the studied wells are the Ca–Mg–Cl and Na–Cl, according to the order of their dominance, with an average sodium/potassium around 40%, 35% carbonate, 70% chloride, 40% sulfate, 60% calcium/magnesium, while calcium reach more than 80% but the wells plotted ranges from less than 20% up to more than 80% because part of the wells pumping from the basalt aquifer and some wells from the highly karstic A7/B2 aquifer and some wells from both aquifers as both aquifers hydrogeological connected in the eastern part of the study area see the cross-section. The range of carbonate concentration among the wells is high again due to the different types of aquifers, the basaltic wells show very low concentration, but the karstic A7B2 wells show higher concentration. The low solubility of minerals through the limestone in the A7/B2 aquifer indicates the major ions (+/-) concentration must be of a natural origin. However, it was observed from the plots that calcium does not exceed sodium and potassium and that chloride-sulfate exceeds other ions, suggesting an anthropogenic source (Sarikhani et al. 2014) which is mainly in this study area due to agricultural activities. Moreover, as the plot shows, the high sulfate concentration indicates an anthropogenic source for the wells classified in the piper diagram above the average sulfated plotted line of the study area wells 40% (more than 65% of the wells plotted above the 40% sulfate average concentration line). The combination of Na–HCO_3 and Ca–Mg–HCO_3 is mainly the result of precipitation water infiltration, IRF, and anthropogenic activity (Gazal 2021b).

Correlation Coefficient Matrix of Major Water Parameters in the Study Area

The correlation between the water parameters measured, such as nitrate and TDS, and sulfate, indicates a strong correlation which can be due to the use of commercial chemical fertilizers used in AZB, which consist mainly of ammonium sulfate compositions. TDS was found to correlate strongly with major cations, magnesium, calcium, sodium (r=0.895, r=0.86, r=0.896, respectively) and moderately acceptable correlation with potassium (r=0.4). TDS correlate strongly with the anions, chloride, sulfate, nitrate, and phosphate (r=0.9496, r=0.863, r=0.6, r=0.6 respectively). Sodium concentrations showed a very good correlation with chloride (r=0.89679) indicating that these ions have been derived from the same sources. calcium and magnesium concentrations showed a very good correlation (r=0.85) indicating the presence of the same source of Ca and Mg, from the dissolution of calcite and dolomite the main minerals in the karstic aquifer's geological formations in the study area A7/B2 (Gazal 2021 b).

Investigation Results of the Natural Protectiveness Potential of Groundwater

1. Based on the hydrogeology, geological studies, fault density, thickness, and texture of the soil layer, the area is considered almost naturally unprotected from the intensive agricultural activity on the surface.
2. Besides studying the geological formations and the soil properties, the climatic data analyses show the fragility of this area in the view of low precipitation and the general characteristic of arid areas, which conclude the natural replenishment of the deterioration groundwater aquifer can be very difficult.
3. The geological and drainage lineament density were studied and the GWR was calculated which indicated a good rate of precipitation is percolating through the vadose zone (infiltrating) to reach the groundwater but the amount is very low as the precipitation rate is low in the study area.
4. The investigations of the hydraulic conductivity (HC) values of the studied area hydrogeological formations show a high HC of these formations.
5. The main climate class in the study area is arid, as well as the very small area in the western part of the case study area is warm-temperate according to the Köppen-Geiger climate classification map modified after Rubel et al. (2017).
6. The mean GWR in the study area is 14 mm/ year, the maximum 42.6 mm/year, the minimum is zero in the south part as the outcropping formation there is the B3 which is aquitard.

7. The calculated geological lineament density in the study area is considered to be high density based on Muthumaniraja et al. (2019) classification. The mean density of geological lineaments is 0.5214 km/km^2 with a standard deviation of 0.4015 and the coefficient of variation is 0.77.

The Overlaying Modeling Techniques Results

In this section, the results of the three overlying modeling steps are described and discussed as follows:

1. Investigate the second component of the GRA (the intrinsic vulnerability model results): (Figure 12) which is the longest step in this section and mostly aims to present the capability of different techniques in groundwater intrinsic vulnerability mapping for groundwater management against contamination in an arid agricultural area. The two DRASTIC approaches were implemented in this study and the results of the agricultural DRASTIC approaches were presented and discussed while the same steps were also applied to implement the ordinary DRASTIC approach and the results of this approach are also presented in the tables and figures. While a former COP model results were utilized aiming to do a comparison with the DRASTIC approaches (ordinary and agricultural DRASTIC) and figure out which model can be applicable and representative in studying the intrinsic vulnerability in arid areas.
2. Create a land use thematic layer to simulate the possible Nitrate contamination load. While the suggested simple approach in this part aims to
3. Creating the Nitrate contamination risk map by combining the land use thematic layer created in the second step with the most representative intrinsic vulnerability map created in the first step. While according to the intrinsic vulnerability scheme in this study (Figure 12) the number of the instinct vulnerability maps is 16 plus the COP intrinsic map.

Vulnerability Model to Investigate the Second Component of the GRA (Intrinsic Vulnerability) Results and Discussion

Three main models were used here, Agricultural DRASTIC, Ordinary DRASTIC, and the former COP vulnerability index model. The two DRASTIC approaches were performed twice, the first scenario without changing the recharge parameter rating, and in the second scenario, the recharge parameter rating was modified by multiplying it by 100 to simulate the effect of the intensive irrigation activities as explained in the methodology chapter.

Result of DRASTIC Parameters and the DRASTIC Indexes Vulnerability Map (DIVM)

Table 9. Range and rating of the Agricultural DRASTIC parameters for the case study area

Depth to water table (D)				
ranges (m)	Aarea	Area (%)	Rating (Dr)	Index*
> 39	3375.652	99.623	1	5
30-39	7.074852	0.2088	2	10
20-30	3.993012	0.1178	3	15
12-20	0.877721	0.0259	5	25
6-12	0.453041	0.0134	7	35
2-6	0.187616	5.54E-03	9	45
0-2	0.18907	5.58E-03	10	50
Weight (Dw) = 5				

Topography/ slope				
Range of slope %	area	Area (%)	Rating (Tr)	Index
0-2	1528.515	36.5938	10	30
2-6	1657.062	39.6713	9	27
6-12	658.3575	15.7616	5	15
12-18	265.9471	6.367	3	9
>18	67.10034	1.6064	1	3
Weight (Tw) = 3				

Aquifer	Area (Km²)	Area (%)	Aquifer media (A)		Hydraulic conductivity (C)	
			Rating (Ar)	Index	Rating (Cr)	Index
A1-A6	228.6716	6.92643	6	18	8	16
B3	263.7302	7.98835	2	6	1	2
B4	20.78728	0.62964	6	18	8	16
Basalt	1367.194	41.4121	9	27	10	20
Kurnub	0.30230	9.16E-03	8	24	10	20
A7-B2	1420.75	43.0343	10	30	10	20
			Weight (Aw) = 3		Weight (Cw) =2	

Recharge (R) according to (Aller 1987)				
Range (m/year)	Area	Area (%)	Rating (Rr)	Index
0.0 -0.05	3329.9	100	1	4
Recharge (R´) suggested ratings				
Range (mm/year)	Area	Area (%)	Rating (Rr)	Index
> or 25	88.59091	2.6605	9	36
18-25	1067.725	32.0646	8	32
10-18	968.2032	29.0759	6	24
5-10	913.9751	27.4474	3	12
0-5	291.4244	8.7517	1	4
Weight (Rw) =4				

	Soil media (S)			
soil texture	Area (Km²)	Area (%)	Rating (Sr)	Index
Clay	133.51	4.01	2	10
Silty clay	704.67	21.18	2.5	12.5
Silty clay loam	1759.86	52.89	3	15
Silty loam	482.85	14.51	4	20
loam	246.48	7.41	5	25
Weight (Sw) = 5				

Soil Thickness in the study area	area (Km2)	Area (%)
0.5 - 1 m	2962.55	89.3095
> 1 m	118.2202	3.5639
< 0.5 m	236.4022	7.1266

According to original modified complex geological map not the simple hydrogeological map		Impact to vadose zone (I)				
geological formation	Age (Epoch)	Aquifer I.C*	Area (Km2)	Area (%)	Rating (Ir)	Index
Na'ur limestone	Cenomanian-Turonian	A1-6	95.75312	2.878	6	24
Wadi As sir limestone	Coniacian	A7	370.642	11.1401	10	40
Wadi Umm Ghudran	Santonian	B1	148.376	4.4596	6	24
Amman silicified limestone	Campanian to Maastrichtian	B2	367.7481	11.0531	10	40
Muwaqqar Chalk Marl	Paleocene-Eocene	B3	10.47362	0.3148	1	4
Umm Rijam (chert - limestone)	Eocene	B4	141.8831	4.2645	6	24
Basalt**	Oligocene to Holocene	Basalt	1024.991	30.8072	9	36
Kurnob Sandstone	Lower Cretaceous (Alpian, Aptian, Neocomian)	K	0.769653	0.0231	8	32
Alluvium and Wadi Sediments	Oligocene to Holocene	Recent	1166.475	35.0597	3	12
			Weight (Iw) = 4			

*: Index is the scoring index by applying the DRASTIC linear equation number 2 (the agricultural DRASTIC approach), **: Basalt in the study area from several basaltic formations, I.C: Aquifer Identifier codes, the same Range and rating parameters used in the second DRASTIC approach (Ordinary) but applying DRASTIC linear equation number 3, so to calculate the Index the parameters weight have to be referred to the equation number 3, see section 2.5.

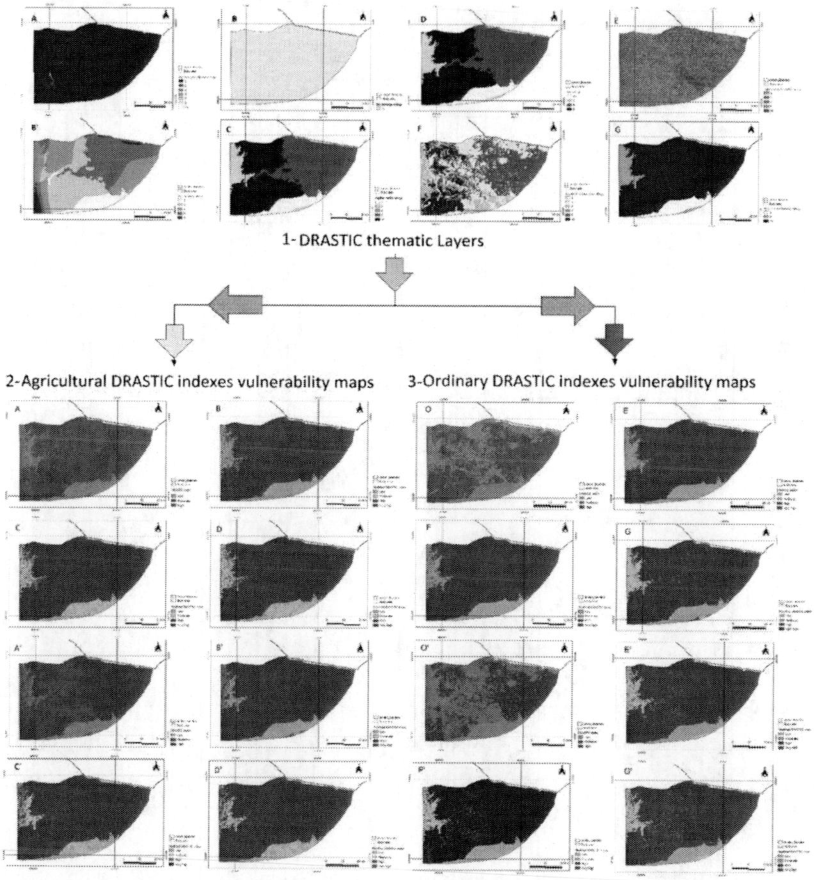

Figure 15. 1) DRASTIC thematic Layers: (A) Depth to water, (B) Net recharge, (C) Aquifer media, (D) Soil, (E) Topography (slope%), (F) Impact of vadose zone, (G) Hydraulic conductivity. (B') Net recharge with a suggested rating.
2) Agricultural DRASTIC indexes vulnerability maps: (A) DRASTIC Vulnerability map A (first scenario), (B) Modified DRASTIC vulnerability map, (C)Modified DRASTIC using extracted 6000 random points, (D) Modified DRASTIC using the extracted 9000points, (the Second scenario with suggested R ratings):
(A') DRASTIC Vulnerability map, (B') Modified DRASTIC, (C') Modified DRASTIC using the extracted 60000 points, (D') Modified DRASTIC using the extracted 900000 points.3) Ordinary DRASTIC indexes vulnerability maps:
(O) DRASTIC Vulnerability map (first scenario), (E) Modified DRASTIC vulnerability map, (F)Modified DRASTIC using extracted 6000 random points, (G) Modified DRASTIC using the extracted 9000points, (the Second scenario with suggested R ratings): (O') DRASTIC Vulnerability map, (E') Modified DRASTIC, (F') Modified DRASTIC using the extracted 60000 points, (G') Modified DRASTIC using the extracted 900000 points.

Table 10. The statistical summary of all the created DVIMs and the percentages of areas covered by the assigned vulnerability classes

MODEL	DRASTIC Qualitative category		Very high	High	Moderate	Low	A statistical summary of the Vulnerability maps was obtained from the raster statistics summary of each map.					
	Map	Range of DRASTIC Indexes	>200	140-200	100-140	1-100		Mean	Minimum	Maximum	S.D*	CV*
AGRICULTURAL	A	(%)	0	12.6401	71.7061	15.6537	Before R*	120.55386	39	171	19.332661	0.16037
		Area *	0	416.7262	2364.04	516.08	After R	1.9698641	1	3	0.5310653	0.26959
	B	(%)	40.4712	45.4974	10.1875	3.8439	Before R	178.94326	38.54239	224.4368	32.155637	0.1797
		Area	1336.077	1502.004	336.32	126.9	After R	3.2296841	1	4	0.7696764	0.23831
	C	(%)	40.7954	45.8052	9.889	3.5104	After R	3.238886	1	4	0.7682529	0.2372
		Area	1344.656	1509.781	325.951	115.705	Before R	181.4108	38.17675	228.8616	33.192957	0.18297
	D	(%)	40.5432	45.4866	10.2839	3.6864	After R	3.228896	1	4	0.7767558	0.24056
		Area	1336.342	1499.28	338.966	121.506	Before R	179.006	38.52064	226.4578	32.276059	0.18031
	A'	(%)	0	55.6655	33.9722	10.4671	After R	2.4546267	1	3	0.6729662	0.27416
		Area	0	1833.39	1118.9	344.743	Before R	138.48848	39	192	24.751174	0.17872
	B'	(%)	32.4107	54.8449	7.3785	5.3659	After R	3.1480977	1	4	0.7698346	0.24454
		Area	1068.592	1808.255	243.272	176.916	Before R	177.81262	37.0173	223.9578	33.394612	0.18781
	C'	(%)	36.2841	50.6752	7.6414	5.3993	After R	3.178471	1	4	0.7895548	0.24841
		Area	1195.958	1670.302	251.869	177.965	Before R	178.9076	36.39834	226.8094	34.75371	0.19426

MODEL	DRASTIC Qualitative category		Very high	High	Moderate	Low	A statistical summary of the Vulnerability maps was obtained from the raster statistics summary of each map.					
	Map	Range of DRASTIC Indexes	>200	140-200	100-140	1-100		Mean	Minimum	Maximum	S.D*	CV*
AGRI-CUL-TURE (Cont.)	D'	(%)	32.3902	54.558	8.6035	4.4483	After R	3.148926	1	4	0.7521341	0.23885
		Area	1057.611	1798.284	283.58	146.619	Before R	177.6217	37.02233	223.9517	33.602388	0.18918
ORDINARY	O	(%)	0	0.000196	0.590544	0.40926	After R*	1.5909367	1	3	0.49206	0.30929
		Area		0.647201	1946.935	1349.267	Before R	110.52887	30	170	22.51896	0.203738
	E	(%)	0.074298	0.779633	0.066754	0.079316	After R	2.84980883	1	4	0.654943	0.22982
		Area	244.9481	2570.331	220.0782	261.4917	Before R	164.716141	31.18456	207.0496	36.06777	0.218969
	F	(%)	0.102751	0.755768	0.060282	0.081199	After R	2.88144646	1	4	0.682568	0.236884
		Area	338.6573	2490.94	198.6854	267.625	Before R	167.317097	30.99712	209.2608	36.992	0.221089
	G	(%)	0.074393	0.780616	0.068999	0.075992	After R	2.8532436	1	4	0.650311	0.22792
		Area	245.2759	2573.717	227.4931	250.5488	Before R	164.990342	31.16264	211.0975	35.88672	0.217508
	O'	(%)	0	0.473492	0.405473	0.121035	After R	2.35628906	1	3	0.684013	0.290292
		Area		1561.118	1336.859	399.057	Before R	128.5163	30	183	27.98408	0.217747
	E'	(%)	0.011902	0.833308	0.075648	0.079142	After R	2.77778692	1	4	0.595435	0.214356
		Area	39.25601	2748.416	249.5011	261.0271	Before R	163.016696	30.10293	208.5107	36.1955	0.222035
	F'	(%)	0.015971	0.830388	0.068079	0.085562	After R	2.77665483	1	4	0.61339273	0.220911
		Area	52.67209	2738.578	224.5225	282.1781	Before R	163.941765	29.746956	210.53909	37.4744613	0.228584
	G'	(%)	0.011686	0.833606	0.075216	0.079492	After R	2.77748546	1	4	0.59611959	0.214626
		Area	38.5289	2748.428	247.9883	262.0893	Before R	162.890783	30.112646	208.48181	36.248837	0.222535

After R*: after reclassifying the vulnerability map according to the DRASTIC qualitative category (1: Low 2: Moderate 3: High 4: Very high), CV*: coefficient of variation, S.D*: stands for standard deviation, Area* in (km^2).

Ranges and ratings of the DRASTIC parameters in the study area were listed in (Table 9). (Figure 15) Displayed the seven rated DRASTIC parameter maps used to calculate the DIVM. The initial agricultural DIVM of both scenarios (A and A´) were obtained using the seven hydrogeological DRASTIC data layers (the seven DRASTIC parameters), according to the agricultural DRASTIC governing equation-2 and processed by the GIS software environment. In the second DRASTIC approach, the initial DIVM of both scenarios (O and O´) were obtained by applying the ordinary DRASTIC model governing equation 3. To understand the created DIVM values, a representation method must be chosen that can reveal the vulnerability of the study area in an acceptable visualization manner and at the same time allow for comparability between different areas (Aller et al. 1987). Therefore, these values were reclassified into four classes using the Jenks natural breaks classification. All the created DIVMs of the study created in this research were classified into four groups according to DRASTIC Indexes scorings (DIS) as low, moderate, high, and very high ((DIS<100), (DIS 100-140), (DIS 140-200) and (DIS>200) respectively). In the Agricultural DRASTIC approach, the range of DIS for both initial vulnerability maps in the first scenario (A) ranges from 39 to 171, and ranges from 139-to 192 in the second scenario (A´) While in the Ordinary DRASTIC approach, the range of the DIS for both initial vulnerability maps in the first scenario (O) ranges from 30 to170, and ranges from 30 to 183 in the second scenario (O´). (Table 10) presents all the created intrinsic vulnerability maps DIS values and the covering percentage of each vulnerability class in the study area.

DRASTIC Statistical and Sensitivity Analyses Result and Discussions
Three statistical analyses were performed in this section, statistical analyses of the DRASTIC rated parameters (SADP), map removal sensitivity analyses (MRSA), and single parameter sensitivity analyses (SDAP), the results and discussion for these analyses are as follows.

Statistical Analyses of DRASTIC Rated Parameters (SADP)
The statistical summary of the seven rated parameter maps used to create the DVIM is provided in (Table 11). The uppermost risk of contamination of groundwater in the study area originates from the hydraulic conductivity parameter (C) (mean value is 9.1). Then the high risk in the study area is caused by the Aquifer media parameter map (A) (mean value is 8.64) and the Topography parameter map (T) (mean value is 8.22). The impact of the vadose zone parameter map (I) (mean value is 6.74) the soil media parameter map (S)

(mean value is 3.14) imply moderate risks of contamination, while depth to groundwater parameter map (D) and the net recharge parameter map (R) impose a very low risk of contamination of groundwater (mean value is 1). While the recharge parameter map (R´) was used to compute the DVIM in the second approach by modifying the recharge scale, not the original recharge rating scale provided by (Aller et al. 1987) see the ranges of (R´) in the (Table 9), the net recharge parameter map (R´) on the second approach has a moderate influence, not a very low as in the first approach (mean value is 5.5).

Table 11. Statistical summary of the DRASTIC parameter maps obtained from the raster statistics summary of each parameter

	Mean	Minimum	Maximum	SD	CV
D	1.008456844	1	10	0.157833866	0.156510283
R	1	1	1	0	0
R´	5.460099765	1	9	2.426668486	0.444436657
A	8.640746545	2	10	2.220990709	0.2570369
S	3.144699878	2	5	0.714368179	0.227165773
T	8.218790147	1	10	2.328928045	0.283366287
I	6.744893275	1	10	3.000613286	0.44487187
C	9.129945878	1	10	2.452714538	0.268645025

S.D. stands for standard deviation and CV for the coefficient of variation, (R´): used a modified rating scale, not the original rating.

(R´) and (I) are highly variable among the DRASTIC rated parameters with the coefficient of variation (CV=0.44), while T, C, A, and S are moderately variable (CV are 0.283 and 0.268, 0.257, 0.227 respectively). D and R are the least variable parameter (CV are 0.1 and 0 respectively). The D and R parameters in the study area mainly got a scoring rate of one because the depth to water in the study area is mostly plotted in the lowest risk range of the depth ranges while the recharge despite the good percentage to recharge to the precipitation (10%) but the amount of recharge very low which indicated to be plotted in the lowest risk range of the recharge DRASTIC ranges and ratings scale (Figure 3-13). (Table 12) provided the statistical summary of the seven used rated parameters calculated by extracting random points files within the study area where each point have the values of those parameters maps after extracting those values by the generated random points layers function in the GIS, the attribute tables of those points files were exported and statistical analyses were done for both points layers (the first layer have 610863 random points, and the second layers have 990261 random points), according to the tow points layers statistic summary it's obvious that when the points number increased the statistic become very close to the summary of the statistical summary obtained from the raster of each parameter map.

Table 12. Statistical summary of DRASTIC rated parameters, calculated in excel sheet, from a randomly generated point within the study area

	Statistical summary of DRASTIC rated parameters created by the statistical analyses of the 610863 random points								Statistical summary of DRASTIC rated parameters created by the statistical analyses of the 990261random points							
	D	R	R′	A	S	T	I	C	D	R	R′	A	S	T	I	C
Average	1.00396	1	6.26115	9.04671	2.97714	8.16858	6.86044	9.77756	1.00746	1	5.43499	8.64209	3.1454	8.21258	6.76713	9.08331
S.d	0.1407	0	2.02055	1.1745	0.53432	2.39218	2.96536	0.629	0.16045	0	2.4248	2.21501	0.7172	2.33362	2.90784	2.5373
Max	10	1	9	10	5	10	10	10	10	1	9	10	5	10	10	10
Min	1	1	0	6	2	0	1	8	1	1	1	2	2	1	1	1
Cv	0.140143	0	0.322712	0.12982	0.179473	0.292851	0.432241	0.064312	0.159258	0	0.44614	0.256304	0.228	0.284152	0.4297	0.279332
Skw	49.12425	####	-0.73736	-1.7382	0.702349	-1.38862	-0.36408	-2.4731	34.36345	#DIV/0!	-0.3641	-2.18608	1.2094	-1.41274	-0.31161	-2.76007
Median	1	1	6	9	3	9	9	10	1	1	6	9	3	9	8.91691	10
Quartile minimum	1	1	0	6	2	0	1	8	1	1	1	2	2	1	1	1
Quartile first 25%	1	1	6	9	2.5	9	3	10	1	1	3	9	2.5	9	3	10
Quartile median 50%	1	1	6	9	3	9	9	10	1	1	6	9	3	9	8.91691	10
Quartile third	1	1	8	10	3	10	9	10	1	1	8	10	3	10	9	10
Q-maximum	10	1	9	10	5	10	10	10	10	1	9	10	5	10	10	10

Max. maximum, min. minimum, Skw. Skewness, S.D. stands for standard deviation and CV for the coefficient of variation, R′ used a suggested rating scale, not the original rating.

Table 13. The correlation analysis summary (correlation matrix) between seven DRASTIC rated parameters by two scenarios with and without changing the recharge ratings

	A correlation matrix created from 610863random points cover the study area								A correlation matrix created from 990261 random points cover the study area							
		D	R	A	S	T	I	C		D	R	A	S	T	I	C
first scenario	D	1							D	1						
	R	#DIV/0!	1						R	#DIV/0!	1					
	A	-0.0539405	#DIV/0!	1					A	0.015887	#DIV/0!	1				
	S	-0.0073442	#DIV/0!	0.2227645	1				S	-0.02058	#DIV/0!	-0.28441	1			
	T	-0.0109012	#DIV/0!	0.3189129	0.4722253	1			T	-0.01438	#DIV/0!	-0.00712	0.371084	1		
	I	0.0242609	#DIV/0!	-0.013621	0.069906	-0.031002	1		I	-0.03321	#DIV/0!	0.249573	-0.03357	-0.12039	1	
	C	-0.0617378	#DIV/0!	0.9176685	0.4473869	0.5016214	0.0096395	1	C	0.011524	#DIV/0!	0.943653	-0.32458	-0.01671	0.250135	1
the second scenario (R′)		D	R′	A	S	T	I	C		D	R′	A	S	T	I	C
	D	1							D	1						
	R′	-0.0508234	1						R′	-0.03832	1					
	A	-0.0539405	-0.086552	1					A	0.015887	0.409222	1				
	S	-0.0073442	0.3346263	0.2227645	1				S	-0.02058	-0.20056	-0.28441	1			
	T	-0.0109012	0.2437691	0.3189129	0.4722253	1			T	-0.01438	0.058538	-0.00712	0.371084	1		
	I	0.0242609	0.0311634	-0.013621	0.069906	-0.031002	1		I	-0.03321	0.206084	0.249573	-0.03357	-0.12039	1	
	C	-0.0617378	0.2075532	0.9176685	0.4473869	0.5016214	0.0096395	1	C	0.011524	0.560525	0.943653	-0.32458	-0.01671	0.250135	1

The summary of the correlation analysis result is provided in (Table 13), The analyses were done by creating random points using Qgis algorithmic function of creating random points layers with a given number of random points, all the points for sure within the study area extent, the number of the points created were (610863 points) of the first trial, then on the second trial other points layer of about (990261 points) were created to compare and for precision results, multiple values from the DRASTIC parameters maps were extracted to each point in those points layers. the original numbers of the generated points were a bit more than those numbers, but null data points were removed.

The attribute tables of the layers of the random points were used to perform the correlation analyses. first, the correlation used the DRASTIC rated parameters when R was rated by the original recharge rating scale, then the correlations analyses were done using the same rated parameters but by using a modified rating scale (R´) instead of R. The correlation analysis between the DRASTIC parameters indicated a strong relationship exists between (A) aquifer media and Hydraulic conductivity (C) ($r > 0.9$) this result can be explained by the same origin of those parameters where they were created from the simplified hydrogeological map. In the first correlation analysis from both points layers where R has been used the correlation between R and all the other parameters does not exist because the value of R is constant and equal to 1. However, in the second analysis, the correlation between R´ and the other DRASTIC parameters exist but show a weak correlation. Due to low correlations between the DRASTIC-rated parameters at (a 95% confidence level), these DRASTIC-rated parameters in the study region were generally considered to be independent.

Map Removal Sensitivity Analysis (MRSA)

The results of these analyses calculated by removing one or more DRASTIC layers at a time from the initial created vulnerability map DVIM for both scenarios are shown in (Table 14) the statistical measures of MRSA of the DVIM for the removal of multiple parameters at once are summarized. In carrying out this MRSA, two or more DRASTIC parameter layers were removed, the DVIM was computed, and the corresponding statistical measures of the variation index from the initial DVIM were calculated. For both scenarios of DRASTIC implementation in this study, it can be noticed from the table of MRSA that with increasing the number of the removed parameters, the variation index does increase, which illustrated the significance of using all seven parameters, otherwise the computed DVIM would be sensitively influenced. (Table 15) presents the results of the MRSA applied to the ordinary DRASTIC approach and the same results repeated by the increasing trend of the variation index by increasing the number of the removed parameters thus indicating the significance of using all the seven DRASTIC parameters, otherwise, the computed ordinary DVIM would be sensitively influenced.

Single-Parameter Sensitivity Analysis (SPSA)

According to these analyses as it can be understood from (Table 16) the agricultural DVIM seems to be most sensitive to the hydraulic conductivity parameter (C) in the first scenario, as it showed a clear high variation of the vulnerability index, with a mean variation index=2.362%. While in the second scenario the removal of the depth parameter (D) showed the highest variation, in general, the agricultural DVIM showed a noted sensitivity to removing any of its seven components in the study area in both scenarios. (Table 17) presents the SPSA results performed for the ordinary DRASTIC which indicated that the ordinary DVIM is most sensitive to the impact to the vadose zone (I) in the first scenario, as it showed the highest variation of the vulnerability index, with mean variation index =2.149%. while in the second scenario the removal of the Topography parameter (S) showed the highest variation, and from the rest results, the ordinary DVIM showed a noted sensitivity of removing any of its seven parameters in both scenarios.

Conclusively, upon both sensitivity analyses results (MRSA) and (SPSA), a significant variation in the DVIM assessment is expected if a lower number of DRASTIC seven parameters have been used. And this conclusion is the aim of implementing the sensitivity analyses according to studies that discussed the possibilities to reduce the parameters of this model to reduce the cost (McLay et al. 2001) as discussed before these sensitivity analyses according to several studies is essential to examine the need to not reduce the number of this model parameters by showing the variations in vulnerability assessment in case of removing a parameter from the DRASTIC calculations, but (Napolitano & Fabbri 1996) discussed the criticism against the DRASTIC parallel to the success of this model, and they introduced a methodology to estimate the real weight of each DRASTIC parameters, which easy approach to modify the DVIM instead of trying to exclude one or more of the DRASTIC parameters by comparing the variation occurs while removing the parameters from the DVIM.

Table 14. Statistics of the one or more map removal sensitivity analysis (Agricultural DRASTIC approach)

Statistics of the map removal sensitivity analysis/ Agricultural DRASTIC first scenario DVIM*						Statistics of the map removal sensitivity analysis/ Agricultural DRASTIC second scenario with suggested (R′)					
Variation index (%)						Variation index (%)					
Parameters used	Mean	Minimum	Maximum	S.D	CV	Parameters used	Mean	Minimum	Maximum	S.D	CV
D, R, S, T, I and C	1.361	0.009	3.399	0.472	34.671	D, R′, S, T, I and C	0.897	0.000	2.867	0.479	53.373
D, R, S, T, and I	2.152	0.000	5.862	0.969	45.006	D, R′, S, T, and I	1.229	0.000	4.816	1.100	89.457
R, S, T, and I	2.234	0.000	7.813	1.637	73.276	R′, S, T, and I	2.074	0.000	7.984	1.642	79.184
R, S and T	2.657	0.000	9.778	2.001	75.31	R′, S and T	2.110	0.000	11.560	1.880	89.067
R and S	6.088	0.016	9.247	1.641	26.95	R and S	5.721	0.000	12.577	2.891	50.532
R	10.864	4.530	11.814	0.673	6.197	R′	5.321	0.000	16.051	2.721	51.136
A	8.165	0.054	20.396	2.831	34.671	I	7.709	0.000	31.691	3.945	51.177
T	7.539	0.001	30.830	4.608	61.123	T	5.600	0.000	30.524	4.599	82.135
D	9.985	0.099	17.515	0.920	9.217	A	5.381	0.000	17.203	2.872	53.373
S	3.074	0.000	24.507	2.958	96.235	D	10.499	0.014	23.101	0.975	9.285
I	9.437	0.000	36.562	5.441	57.655	S	4.281	0.000	26.467	2.730	63.781
C	5.714	0.000	12.500	3.213	118.377	C	4.280	0.000	12.646	3.048	133.708

*The initial overall vulnerability index used in the calculation here for the first scenario is for sure the DVIM (A), but in the second scenario is (A′).

Table 15. Statistics of the one or more map removal sensitivity analysis (Ordinary DRASTIC approach)

Statistics of the map removal sensitivity analysis/ Ordinary DRASTIC first scenario DVIM*						Statistics of the map removal sensitivity analysis/ Ordinary DRASTIC second scenario with suggested (R′)					
Variation index (%)						Variation index (%)					
Parameters used	Mean	Minimum	Maximum	S.D	CV	Parameters used	Mean	Minimum	Maximum	S.D	CV
D, R, S, T, I and C	1.606	0.003	3.626	0.644	40.086	D, R′, S, T, I and C	1.123	0.000	9.525	0.723	64.391
D, R, S, T, and I	4.218	0.006	7.813	1.422	33.706	D, R′, S, T, and I	2.901	0.007	9.726	1.272	43.840
R, S, T, and I	3.139	0.000	8.203	1.834	58.446	R′, S, T, and I	1.740	0.000	13.190	1.634	93.918
R, S and T	8.381	0.002	11.866	2.083	24.850	R′, S and T	4.765	0.000	11.882	2.068	43.396
R and S	8.381	0.002	11.866	2.083	24.850	R and S	3.682	0.000	10.563	2.359	64.081
R	10.448	1.242	11.742	1.147	10.978	R′	5.776	0.000	21.453	3.035	52.545
A	9.634	0.018	21.754	3.862	40.086	A	10.056	0.000	19.110	1.575	15.662
T	6.738	0.000	13.486	2.426	36.008	T	9.127	0.007	11.861	2.118	23.207
D	9.467	0.000	20.997	1.477	15.607	D	10.056	0.000	19.110	1.575	15.662
S	8.419	0.007	11.278	2.007	23.832	S	7.691	0.000	13.639	2.359	30.669
I	15.397	0.013	51.312	9.917	64.409	I	12.364	0.000	50.323	8.526	68.954
C	11.580	0.119	20.909	3.403	29.390	C	8.139	0.034	18.098	2.907	35.713

*The initial overall vulnerability index used in the calculation here for the first scenario is for sure the DVIM (O), but in the second scenario is (O′).

Table 16. Statistics of the one map removal sensitivity analysis (Agricultural DRASTIC)

Statistics of MRSA/ for the first scenario Agricultural DVIM*						Statistics of MRSA/ for the second scenario (with suggested R′)					
Removed parameter	Variation index (%)					Removed parameter	Variation index (%)				
	Mean	Minimum	Maximum	S.D	CV		Mean	Minimum	Maximum	S.D	CV
5D	1.6642	0.0165	2.91922	0.1534	9.2174	5D	1.7499	0.00239	3.8502	0.1625	9.285
4R	1.8107	0.7549361	1.96901	0.1122	6.1968	4R′	0.8869	0	2.6752	0.4535	51.136
3A	1.3608	0.00894	3.39939	0.4718	34.6714	3A	0.8969	2.82E-15	2.8672	0.4787	53.373
5S	0.5123	0	4.08453	0.4931	96.2352	5S	0.7135	0	4.4112	0.4551	63.781
3T	1.2566	0.0000885	5.13831	0.768	61.123	3T	0.9333	1.15E-09	5.0874	0.7665	82.135
4I	0.4524	0	2.08333	0.5355	118.3765	4I	1.2848	0	5.2819	0.6575	51.177
2C	2.362	2.3191201	2.38082	0.0128	0.5425	2C	0.38	0	2.1077	0.508	133.708

*The initial overall vulnerability index used in the calculation here for the first scenario is for sure the DVIM (A), but in the second scenario is (A′).

Table 17. Statistics of the one map removal sensitivity analysis (Ordinary DRASTIC)

Statistics of MRSA/ for the first scenario Ordinary DVIM*						Statistics of MRSA/ for the second scenario (with suggested R′)					
Removed parameter	Variation index (%)					Removed parameter	Variation index (%)				
	Mean	Minimum	Maximum	S.D	CV		Mean	Minimum	Maximum	S.D	CV
5D	1.578	0.000	3.499	0.246	15.607	5D	1.713	0.000	11.128	0.506	29.564
4R	0.774	0.000	6.053	0.647	83.629	4R′	1.002	0.000	10.448	0.613	61.190
3A	1.606	0.003	3.626	0.644	40.086	3A	1.123	0.000	9.525	0.723	64.391
5S	1.403	0.001	1.880	0.334	23.832	5S	1.568	0.000	9.525	0.551	35.104
3T	1.123	0.000	2.248	0.404	36.008	3T	13.136	10.012	14.180	0.552	4.202
4I	2.149	1.588	2.318	0.106	4.915	4I	2.098	0.000	8.262	1.439	68.581
2C	1.930	0.020	3.485	0.567	29.390	2C	1.376	0.002	11.003	0.594	43.180

*The initial overall vulnerability index used in the calculation here for the first scenario is for sure the DVIM (O), but in the second scenario is (O′).

DRASTIC Modifications

(Table 18) and (Table 19) summarized the statistical analysis to compare the effective (real) weight that each parameter had in the study area for both scenarios (Napolitano & Fabbri 1996). In both tables, the effective weight was calculated by applying equation 6, but in the first table, the calculations were implemented as the ordinary approach introduced by (Napolitano & Fabbri 1996) to calculate the effective weight directly by applying the equation on the parameters and the initial DVIM rasters, while in (Table 19) the same equation was applied to all the extracted points.

In both tables, the weight value of each parameter used to modify the Agricultural DVIM was calculated from the mean % by multiplying it by the total Agricultural DRASTIC seven parameters weight (26) which is fixed by (Aller et al. 1987) to the agricultural DRASTIC model used in this research. In general, as the tables of the effective weights show the effective weights of the depth to water parameter (D) is less than the original assigned weight this is due to the low variability of this parameter and low-risk contribution to contamination as almost all the study area belongs to the lowest risk DRASTIC (D) range. While it can be concluded here that the real weight for the aquifer media, impact to vadose zone, and hydraulic conductivity increased as these elements reflect the geological formation in the study area which are dominated by karstic and fractured basaltic formations thus have a high influence on the agricultural DVIM assessment. The same steps were applied to modify the ordinary DRASTIC, and the calculated effective parameters weight in (Table 20), and (Table 21). The effective weight for impact to vadose zone (I), aquifer media (A), and hydraulic conductivity (C) are higher than the theoretical weight.

As a result, the effective weight results indicate that the three geological parameters, which have an effective weight higher than the theoretical weight, generally regulate the vulnerability in the region. (Figure 15) presents the updated DVIM of the study area following the statistical modifications made using the effective weights of the seven parameters measured for the agricultural and ordinary DRASTIC approaches and the two applied scenarios in three ways discussed above. The updated DVIM indicates that the case study region is under low, moderate, high, and very high groundwater vulnerability to contamination.

Table 18. Calculations of the DRASTIC parameters effective weight by rasters calculations (Agricultural DRASTIC)

The calculation for the first scenario *									The calculation for the second scenario (with suggested R′)								
parameter	Theoretical		Effective weight (weight (%))						parameter	Theoretical		Effective weight (weight (%))					
	Weight	Weight (%)	Mean	Min.	Max.	S.D	CV	Weight		Weight	Weight (%)	Mean	Min.	Max.	S.D	CV	Weight
D	5	19.231	4.316	3.247	32.154	1.053	24.407	1.122	D	5	19.231	3.799	2.874	38.462	1.085	28.56	0.988
R	4	15.385	3.425	2.339	10.256	0.688	20.081	0.891	R′	4	15.385	15.215	2.614	33.333	5.941	39.05	3.956
A	3	11.538	21.394	5.357	35.714	5.009	23.414	5.562	A	3	11.538	18.584	5	32.432	4.39	23.62	4.832
S	5	19.231	13.423	7.194	40.984	4.218	31.424	3.49	S	5	19.231	11.895	5.848	40.984	4.508	37.9	3.093
T	3	11.538	20.696	2.362	45.455	6.463	31.228	5.381	T	3	11.538	18.266	2.048	45.455	6.371	34.88	4.749
I	4	15.385	21.78	4.706	53.333	8.142	37.382	5.663	I	4	15.385	19.219	3.54	45.977	7.313	38.05	4.997
C	2	7.692	14.966	1.786	23.81	4.193	28.018	3.891	C	2	7.692	13.022	1.667	22.695	3.484	26.75	3.386
	26	100	100					26		26	100	100					26

*The initial overall vulnerability index used in the calculation here for the first scenario is for sure the DVIM (A), but in the second scenario is (A′).

Table 19. Calculated the effective weight of the DRASTIC parameters using the extracted Excel sheets from the DRASTIC rated parameter maps (Agricultural DRASTIC)

	Scenario 1				Scenario 2 with the suggested R′			
	Using the 610863 random points		*Using the 990261 random points*		*Using the 610863 random points*		*Using the 990261 random points*	
	Mean %	Weight	Mean %	Weight	Mean %	Weight	Mean %	Weight
D	4.179724	1.086728	4.308022	1.120086	3.568591	0.927834	3.798113	0.987509
R	3.332674	0.866495	3.422962	0.88997	17.341178	4.508706	15.221225	3.957518
A	22.379538	5.81868	21.329534	5.545679	19.140868	4.976626	18.595448	4.834816
S	12.22706	3.179036	13.41339	3.487482	10.426372	2.710857	11.899057	3.093755
T	19.834183	5.156887	20.691592	5.379814	16.935583	4.403252	18.288559	4.755025
I	21.856717	5.682746	21.922628	5.699883	18.767293	4.879496	19.263064	5.008397
C	16.190104	4.209427	14.911873	3.877087	13.820114	3.59323	12.934534	3.362979
	sum	26	sum	26	sum	26	sum	26

Table 20. Calculations of the DRASTIC parameters effective weight by raster's calculations (Ordinary DRASTIC)

The calculation for the first scenario *									The calculation for the second scenario (with suggested R′)								
parameters	Theoretical		Effective weight (weight (%))						parameters	Theoretical		Effective weight (weight (%))					
	Weight	Weight (%)	Mean	Minimum	Maximum	S.d	Cv	Weight		Weight	Weight (%)	Mean	Minimum	Maximum	S.d	Cv	Weight
D	5	21.74	4.84	3.60	34.97	1.59	4.54	1.11282	D	5	21.74	4.239	3.012	40.984	1.671	39.426	0.975
R	4	17.39	3.84	2.35	13.33	1.16	8.69	0.88351	R′	4	17.39	16.371	2.632	35.294	6.137	37.486	3.765
A	3	13.04	23.35	6.82	36.00	5.16	14.33	5.3702	A	3	13.04	20.085	6.122	32.609	4.446	22.138	4.620
S	2	8.70	6.12	3.01	26.32	2.86	10.89	1.40721	S	2	8.70	5.399	2.424	26.316	2.984	55.269	1.242
T	1	4.35	7.87	0.74	25.64	3.32	12.96	1.81	T	1	4.35	6.899	0.637	25.641	3.312	48.005	1.587
I	5	21.74	29.62	9.26	66.67	10.22	15.33	6.81243	I	5	21.74	26.011	5.155	63.291	9.601	36.912	5.983
C	3	13.04	24.36	3.41	33.80	6.73	19.90	5.60393	C	3	13.04	20.996	3.061	33.803	5.492	26.155	4.829
	23	100	100					**23**		**23**	100	100					**23**

*The initial overall vulnerability index used in the calculation here for the first scenario is for sure the DVIM (O), but in the second scenario is (O′).

Table 21. Calculated the effective weight of the DRASTIC parameters using the extracted Excel sheets from the DRASTIC rated parameter maps (Ordinary DRASTIC)

	Scenario 1				Scenario 2 with the suggested R′			
	Using the 610863 random points *		*Using the 990261 random points* **		*Using the 610863 random points*		*Using the 990261 random points*	
	Mean %	Weight	Mean %	Weight	Mean %	Weight	Mean %	Weight
D	4.50	1.04	4.830	1.111	3.799	0.87	4.24	0.98
R	3.59	0.83	3.839	0.883	18.494	4.25	16.38	3.77
A	24.17	5.56	23.260	5.350	20.418	4.70	20.11	4.62
S	5.30	1.22	6.115	1.406	4.458	1.03	5.41	1.24
T	7.23	1.66	7.871	1.810	6.086	1.40	6.92	1.59
I	28.98	6.67	29.832	6.861	24.644	5.67	26.11	6.01
C	26.22	6.03	24.253	5.578	22.101	5.08	20.83	4.79
	sum	23	sum	23	sum	23	sum	23

*the extracted random points Excel sheet-1 ** the extracted random points Excel sheet-2.

The second scenario by the suggested R′ doesn't offer better visualization of the reality and it was enough to adjust and modify the DVIM by the effective weights. The updated DVIM in both scenarios using the effective weights calculated by the extracted points values shows similar results close to the updated DVIM results by the values of the effective weights calculated directly by applying the effective weights calculation approaches of (Napolitano & Fabbri 1996) on the raster's, especially the second estimations as the number of the points increased (map D and D′ in the agricultural DRASTIC approach) and (map G and map G′ in the Ordinary DRASTIC approach) see (Table 9), and (Figure 15). (Figure 17) show the percentages of each assigned vulnerability class in the study area by each prepared DVIM and the statistical summary of each map. The modification by the effective (real) weight for both agricultural DVIM generated by the two adopted scenarios (with original R and with modified R′), shows a new vulnerability indexes class the very high vulnerability in the modified agricultural DVIM (B, and B′) which wasn't present in the initial agricultural DVIM (A and A′). Also, the same results obtained after the modification by the effective weight for both ordinary DVIM generated by both scenarios, a new vulnerability indexes class the very high vulnerability presents in the modified agricultural DVIM (E and E′) which this very high vulnerability class wasn't present in the initial ordinary DVIM (O and O′). The same high vulnerability class also presents in the other modified DVI maps by the effective weights computed by the extracted point values but applying the same approach used in the raster calculated in the two extracted points excel sheets (excel sheet1 with 610863 points and excel sheet-2 with 990261points). While the percentages of the vulnerability classes assigned after the modification by the real weight created by the excel sheet calculation show almost the same percentages as the vulnerability maps (B, B′, E, and E′) which were modified by the effective weights calculated by the raster calculator, especially the DVI maps (D, D′, G, and G′) than the modified DIV maps (C, C′, F, and F′) because the number of the points extracted in the second excel sheet more than the number of the extracted points in the first excel sheet (990261 points, and 610863 points respectively).

The explanation for the existence of a new vulnerability class (a very high vulnerability class) in all modified DVIMs can be explained by reference to the effective weight tables, which indicates an increase in (I, C, and A) weight as these DRASTIC-parameters have a greater effect on the vulnerability index assessment process in the study area while the weight of the less influencing parameters (D and R) decreased (Table 18, Table 19, Table 20, Table 21). This approach to finding the real DRASTIC-parameters weight make this model more flexible and able to represent the real natural protectiveness of the groundwater aquifer in different areas. For example, in this research, the actual measured weight of parameters D and R is 1.1 and 0.8, while the original theoretical weight for these parameters is 5 and 4 respectively. Although applying the same model and approach to the measurement of real DRASTIC-parameters weight in areas with a higher rate of recharge as the findings of (Napolitano & Fabbri 1996) throughout the study area in Italy gave a higher weight for R than the original theoretical weight given by (Aller et al. 1987).

DRASTIC Validation

The agricultural modified DRASTIC map (B) was selected to represent the study area's intrinsic groundwater vulnerability as the ordinary DRASTIC less estimated the vulnerability of the study area and the Agricultural DRASTIC was designed by (Aller et al. 1987) to assess the vulnerability when the agricultural activities are the main source of the possible contamination. Although the agricultural modified DRASTIC map (B´) by the second scenario (the modified R′ ratings) doesn't give better visualization as mentioned before so the modification by calculating the effective weight was enough in this study. Therefore, the agricultural modified DRASTIC map (B) was validated by studying the matching and correlation between the groundwater vulnerability map (B) and the water quality SCD maps. Besides the comparison with the COP intrinsic vulnerability indexes map (Figure 16) created by (BGR & MWI 2018).

There is a relatively very close probability of distribution according to the coefficient of variation (relative standard deviation) for each parameter spatial distribution map 57%, 61%, 59%,59%,65%, and 59% for the average and maximum nitrate, sulfate, and EC, SCD maps respectively. Besides, the raster's correlation of the three water quality parameters SCD maps also shows strong spatial relative dependents with the raster's correlations values, between nitrate and EC (r=85%), between sulfate and EC (r=90%), nitrate and sulfate (r=78%). This strong correlation indicated a possibility the same resource causing the increase of these parameters' concentrations, which is related to the effects of the large-scale source (non-point pollution sources), which is the agricultural activities, or may contribute to large-scale pollution controlling factors such the weak groundwater natural potential protectiveness in the study area. The high EC values are also related to agricultural activities by the effect of the (IRF). The modified agricultural DVIM corresponds to the continuous distribution maps of the average and maximum concentrations of nitrate, sulfate, and EC elements. The agricultural activity intensity map which was prepared depending on the agricultural intensity data retrieved from (MOA 2020), also comes in parallel to the susceptibility approaches by DRASTIC as the high intensive agricultural activities occur in the north-eastern part of the study area which according to modified agricultural DVIM (B) is a highly vulnerable area, but the high concentration of nitrate, sulfate, and EC is clear in the middle of the study area which is within the high vulnerability zone according to (B) map, the explanation comes through the groundwater movement which moves from high elevation to the lowest and according to the static groundwater level map the groundwater moves from the north-eastern part which has high intensive agricultural activities to the middle part.

According to COP-model, the study area is assigned as the very low to moderate vulnerable areas (Figure 17, Figure 16). This is against the study area's geological nature which is characterized by highly karstification aquifers and fractured high permeable basalt in addition to the failure of this model to coop with the several studies discussed the groundwater quality deterioration and the increasing trend of groundwater contamination and the recommendations to regulator the agricultural activities in this highly vulnerable area.

This underestimation of the groundwater susceptibility in arid low precipitation areas is due to the COP-model govern equation (Vías et al. 2006) which was developed for the regions in Germany characterized by humid climate and high annual precipitation. COP-model by giving high weight to the precipitation in calculating the vulnerability indexes can give a representative vulnerability map for the wet areas but in dry areas, it may have weakness by underestimating the possible highly vulnerable areas, and by creating the COP vulnerability map it wasn't subjected to any statistical modification analyses or validations. The influence of precipitation in vulnerability assessment by the DRASTIC model is presented by the groundwater recharge and it is theoretical weight percentage relative to the rest of the seven elements of this model is 15.3% after finding the actual weight of R the effective weight percent of R became 3.4% (Table 18). Therefore, the DRASTIC model is more appropriate for dry areas to simulate the susceptibility of pollution. While the simple approaches of findings the effective weight of DRASTIC parameters makes it a more flexible model to be used in dry and wet areas. Besides in arid areas even the precipitation is low the groundwater recharge can be increased in some agricultural areas due to excessive IRF (Srivastava 2013) therefore, the COP model isn't suitable to evaluate the natural potential protectiveness of groundwater in arid areas, especially agricultural arid areas.

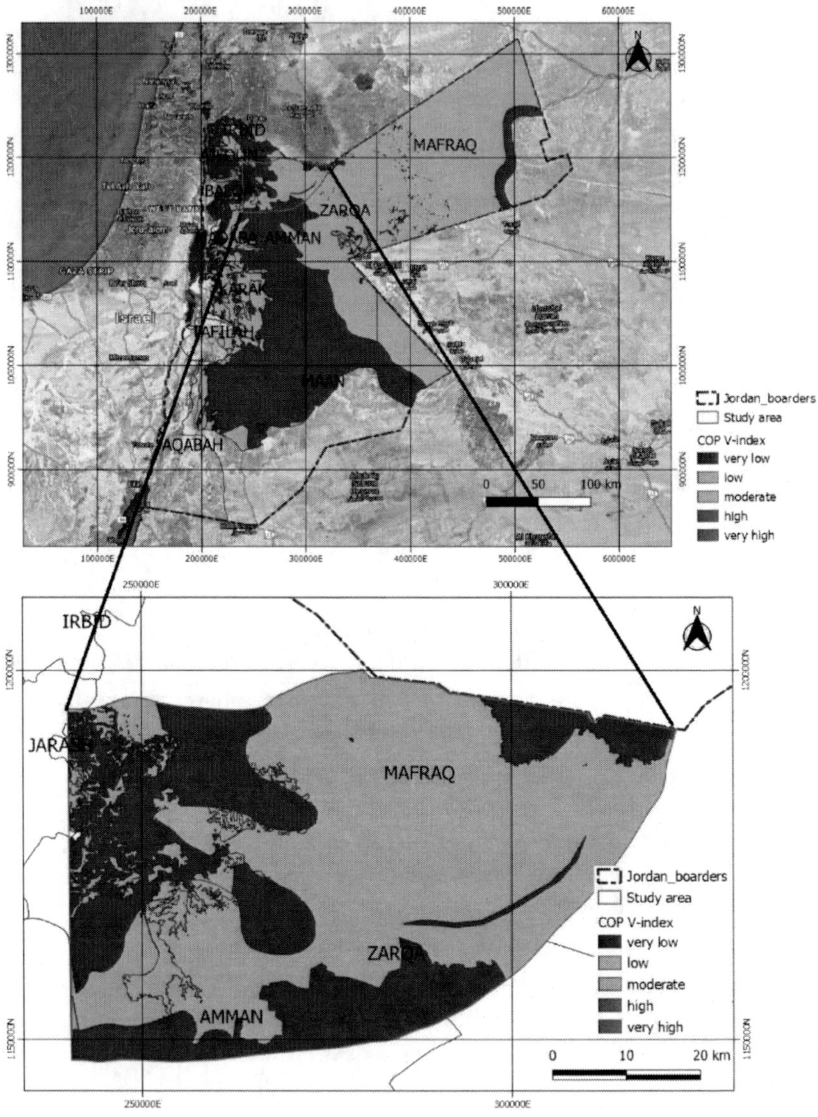

Figure 16. COP vulnerability map retrieved from (BGR & MWI 2018).

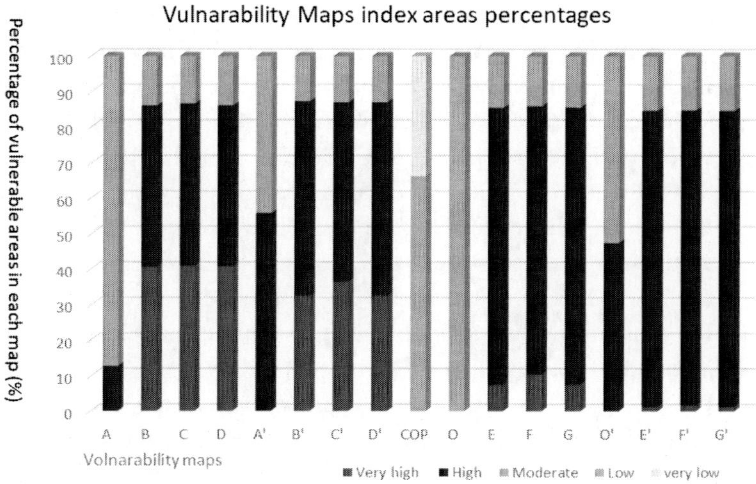

Figure 17. Percentage of vulnerable areas in each approach: (A) DRASTIC Vulnerability map A (first scenario), (B) Modified DRASTIC vulnerability map, (C)Modified DRASTIC using extracted 6000 random points, (D) Modified DRASTIC using the extracted 9000points, (the Second scenario with suggested R ratings): (A') DRASTIC Vulnerability map, (B') Modified DRASTIC, (C') Modified DRASTIC using the extracted 60000 points, (D') Modified DRASTIC using the extracted 900000 points. And the middle column represents the areas percentages assigned vulnerability classes by the COP-Model, (O) DRASTIC Vulnerability map (first scenario), (E) Modified DRASTIC vulnerability map, (F)Modified DRASTIC using extracted 6000 random points, (G) Modified DRASTIC using the extracted 9000points, (the Second scenario with suggested R ratings): (O') DRASTIC Vulnerability map, (E') Modified DRASTIC, (F') Modified DRASTIC using the extracted 60000 points, (G') Modified DRASTIC using the extracted 900000 points.

Groundwater Nitrate Contamination Risk (NCR) Map

This part was created to simplify the use of an enormous amount of spatial data, using overlying modeling techniques, to produce the visualization results of groundwater nitrate contamination risk (NCR) in the study area, summarized below. The results of the three overlying models used in this research are shown in (Figure 18).

These simulations represent as far as possible:

1. The natural groundwater protectiveness by intrinsic vulnerability model, via the agricultural modified (DVIM) DRASTIC vulnerability indexes map (B), which was selected after creating several modified

DVIMs by the two DRASTIC approaches the Ordinary and the Agricultural, and by adopting two implementation scenarios for each DRASTIC approaches and modified all the resulted DVIMs by the statistical approach introduced by (Napolitano & Fabbri 1996) see the whole results of this part in section (4.3.1 above). DVIM (B) indicates that 40% of the study area fall under the very high vulnerability condition (value of the DRASTIC Indexes scorings (DIS) > 200) and about 45% of the study area fall under the high vulnerability condition (value of the DIS is between 140-200), which reflect the fragility of the groundwater natural protectiveness in the study area (Figure 17). While the findings of the GRA investigations suggest that agricultural practices are the key cause of contamination, the natural potential protectiveness of the groundwater by studying the geological and hydrogeological parameters revealed a fragility of this protectiveness.

2. The nitrate potential contamination by a land-use thematic map created via a simple overlying model to simulate the possible nitrate contamination loads. The overly of the three-land use nitrate contamination potential maps indicated parallel to the contamination load investigation that the main source of the nitrate contamination in the study area is related to agricultural activities, where the study area is classified into very high, high, moderate, low, and very high APNC classes in 8.35%, 5.92%, 27.84%, 27.51% and 30.36% of the studied area, respectively (Table 22). While very low, low, moderate, and high UPNC classes were classified in 95.9%, 3.5%, 0.9, and 0.01% of the studied area, respectively. Then very low and low PuPNC classes were classified in 95% and 5% of the studied area, respectively. The PNC map resulted by overlapping the previous three classified land use thematic layers responding to the nitrate contamination factors indicated that the study area was classified into five PNC classes from very low to very high (Figure 18).
3. The nitrate contamination risk (NCR) map created by overlapping the PNC map and the modified agricultural DVIM (B) of equal weight (using the arithmetic average of the input values) indicated the study area assigned a very high, high, moderate, low, very low NCR classes in 2.96%, 25.42%, 51.72%, 16.73%, 3.15% of the studied area, respectively.

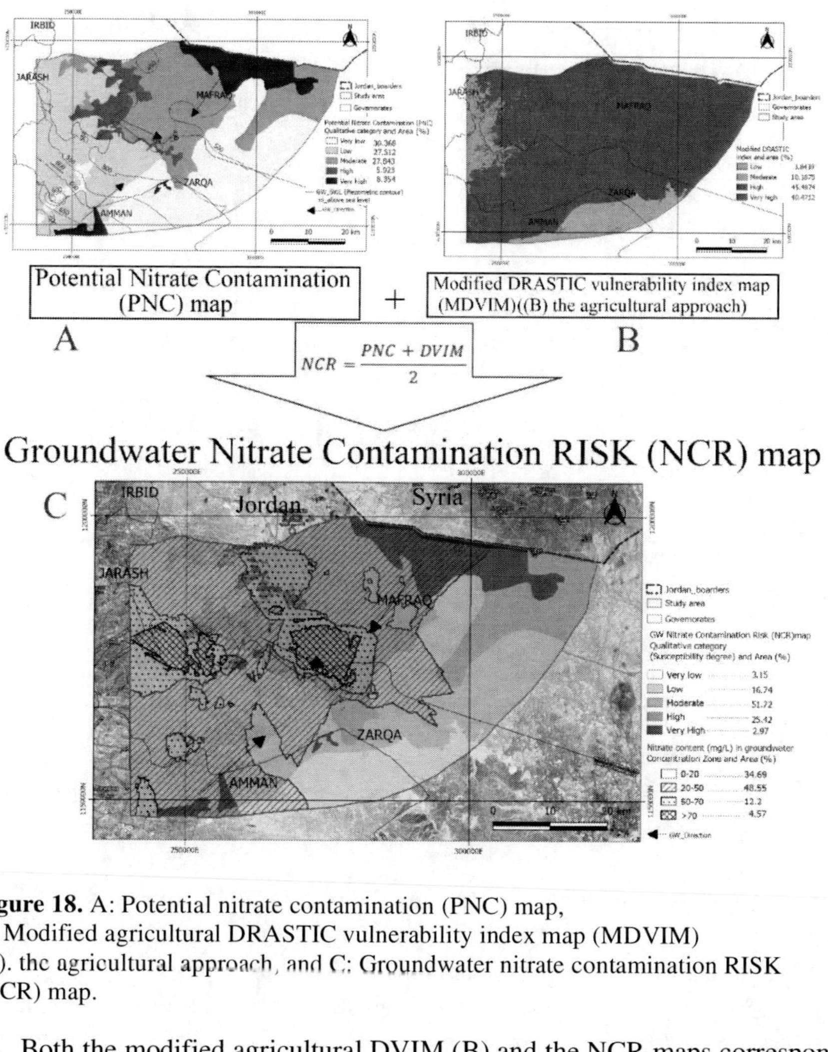

Figure 18. A: Potential nitrate contamination (PNC) map,
B: Modified agricultural DRASTIC vulnerability index map (MDVIM)
(B). the agricultural approach, and C: Groundwater nitrate contamination RISK (NCR) map.

Both the modified agricultural DVIM (B) and the NCR maps correspond to the continuous distribution maps of the average concentrations of nitrate, sulfate, and TDS elements, where it was discussed the matching distribution relation of both nitrate, sulfate, and EC concentration maps as the source of these elements is the agricultural contaminations via the fertilizers used in the study area which include these elements.

Table 22. The three-land use thematic layers responding to the nitrate contamination factors. Classified into five classes according to (Ducci 2018) land use approach

A: Land use classification in the Agricultural Potential Nitrate Contamination.					
Km2	Agricultural activities intensity		Area (%)	Agricultural Potential Nitrate Contamination (APNC)	
277.9706	Agric. > 75%*		8.354084	Very high	
197.0878	Agric. 50-75%		5.923244	High	
926.4287	Agric. 25-50%		27.84274	Moderate	
129.0549	Agric. 10-25%		3.878595	Low	
786.3686	Agric. < 10%		23.63339	Low	
951.5815	No agric., dom. pasture		28.59867	Very low	
58.87007	No agric., other land uses		1.769272	Very low	
B: Population density classes and reclassification in the Urban Potential Nitrate Contamination.			C: Sewer system coverage and reclassification in the Peri-urban Potential Nitrate Contamination.		
Population density per km2	Urban Potential Nitrate Contamination (UPNC)	Area (%)	Sewer system coverage %	Peri-urban Potential Nitrate Contamination (PuPNC)	Area (%)
<5	Very low	95.55	>90	Very low	95
25-May	Low	3	70-90	Low	5
25-250	Moderate	1	50-70	Moderate	
250-1000	High	0.1	25-50	High	
>1000	Very high	0	<25	Very high	

*75% of the area is equipped with intensive farming.

Completely, the matching between the resulted NCR map and the nitrate SCD map (Figure 18) to evaluate the overlying modeling techniques results indicated a strong corresponding matching between the NCR and the nitrate SCD zones. But the highly intensive agricultural activities occur in the north-eastern part of the study area which according to DVIM is a very highly vulnerable area and also categorized as a very high NCR area, but the high concentration of nitrate is clearer in the middle of the study area which is within the high vulnerability zone according to modified DVIM and categorized as high NCR area, not very high. The explanation comes through the groundwater movement which moves from high elevation to the lowest according to Darcy law and by referring to the static groundwater level map (E) (Figure 14) the groundwater moves from the north-eastern part which has the highly intensive agricultural activities to the middle part of the study area. As mentioned in the introduction page number 18 the intrinsic vulnerability map is more applicable to be used for land-use planning because the contamination risk map does not reflect the intrinsic groundwater vulnerability but also the contamination status which is subjected to be changed according to the human activities on the surface. Only the intrinsic vulnerability can also have the same weakness in the case of massive human activities causing removal or disruption of the vadose zone.

Case Study Conclusion

The overlying modeling techniques are used to simulate the intrinsic vulnerability and the specific vulnerability by adding the land use created thematic layer to the modified DRASTIC intrinsic vulnerability map. The DRASTIC model, even though it gives relatively satisfactory results in the evaluation of groundwater intrinsic vulnerability to pollution, cannot be used for the truthful assessment of the groundwater pollution risk in arid highly permeable areas without modifications. Besides, the hydrogeological conditions provided in the original DRASTIC model by (Aller et al. 1987) do not make special provisions for hazardously sensitive, karstic, and fractured basaltic rocks domains. However, the flexibility of the DRASTIC model and its ability to be applied in areas belonging to different environmental characteristics was clear by the ease, efficiency, and simplicity of finding the parameter's real weight in the modification approach.

Using the statistical sensitivity analysis approach to avoid subjectivity associated with the selection of ratings and weights of the seven model parameters conclude that the seven DRASTIC components were essential and significant in the calculations of DVIM. Furthermore, optimizing the weights of the DRASTIC parameters using the aforementioned optimization procedure by (Napolitano & Fabbri 1996) which can be easily achieved using simple statistical approaches was more effective and reliable than changing the ratings of the DRASTIC parameters as suggested in the second scenario (R'). The statistical analyses show that aquifer media (A) and hydraulic conductivity (C) are the most significant parameters which dictate the high vulnerability in the study area. This highlights the importance of finding accurate, comprehensive, and representative information about these parameters which leads to checking the accuracy of the hydraulic conductivity values used in this study.

The correlation between nitrate and sulfate and the EC was high, suggesting that agricultural activities are the main cause of groundwater deterioration in the study area, while the correlation between the modified DIVM (both DRASTIC approaches the ordinary and Agricultural) and the concentration of contaminants indicates a better representation of the fragility of the aquifer than the COP-model. Besides DRASTIC-model, shows the great correspondence between the sensitivity of the area to pollution and the distribution of geological outcropping. One of the key findings of this research is the importance of using a suitable vulnerability model that is appropriate for hydrogeological, climatic, and contamination risk load in the study area. While, referring to the comparison results between COP-model and DRASTIC model its recommended not to use, COP-model for dry areas and to use DRASTIC but in general the application of DRASTIC in any region needs to be adjusted by finding the actual weight of the seven parameters to make the DVIM more reflective of the reality and thereby improve the consistency of the implementation of the model.

Using the contamination-prone maps models showed the ease of stimulating the area's susceptibility to pollution in a visualization map. And the overlying modeling techniques used in this study are very simple to use and require the appropriate management of agricultural areas, such that areas of high vulnerability can be limited to certain types of land use. According to the GRA investigation of the first component (contamination loads), the key sources of contamination in the study area derived from agricultural activities, on this finding the agricultural DRASTIC model was chosen as recommended by (Aller et al. 1987). Also, a comparison of the percentage areas in the vulnerability classes using the modified DRASTIC model via several suggested scenarios, concludes that the modified agricultural DRASTIC approach is more applicable than the Ordinary Drastic and COP model which underestimated the degree of the intrinsic vulnerability in the study area.

The modified DRASTIC results showed the fragility of the groundwater aquifer, especially in the central and north-eastern part of the study area, requiring a serious protection plan. While the modified agricultural DIVM demonstrated that (40%, 45%, 10%, 3%) of the case study area is under very high, high, moderate, and low groundwater vulnerability to contamination respectively. This calls for an urgent plan to control the spread of agricultural activities, especially in the mid-and north-eastern areas, while the southern part of which about 3% of the study area is somehow naturally protected as the outcropping layer is aquitard. While the groundwater NCR indicated the study area, assigned very high, high, moderate, low, very low NCR classes in 2.96%, 25.42%, 51.72%, 16.73%, 3.15% of the studied area, respectively.

The groundwater NCR map was created by merging intrinsic vulnerabilities and possible nitrate contamination load maps, this makes it differs from the created modified DVIM because it includes the evaluation of the nitrate contamination potential. But it represents a particular current groundwater contamination risk associated with the selected nitrate contamination parameter in this study, whereas the current land use might have been changed so that the created groundwater NCR may not be reliable and may require to be modified to represent changes in land use or changes in the selected contamination parameters. Therefore, the modified DVIM is more reliable to be used for future strategic land use planning than the current groundwater NCR.

Ultimately, the main benefits of the approaches proposed in this study for the comprehensive groundwater risk assessment as a whole, including the investigation of GRA components, and the overlaying modeling approaches, are the potential of being implemented on a broad scale, simple availability, and versatility of starting data, even from various sources. Besides, the reliability of the overlying modeling techniques used to create the NCR map (in terms of accordance with the nitrate concentration zones) and the intellectual methodology of the NCR map generated, seems to be useful and have a significant potential for being employed worldwide.

Key Findings

The efforts to incorporate the methodological approaches proposed here would result in a cost-effective solution to maintaining the source of community drinking water and achieving the desired protection of groundwater for future generations. The research can also be used as a method to raise public awareness of groundwater problems in developed countries. The research is intended to establish a holistic pattern for the use of several scientific methods in a systematic and monotonous manner to investigate groundwater risk in support of sustainable groundwater management. The study has shown that the overlay models used to create contamination vulnerability maps are better suited for dry areas with Jordan as an example. The developed integrated overlaying approach is mostly aimed at assessing groundwater vulnerability and evaluating the estimation problem of nitrate contamination in arid agricultural areas. This included the use of the updated version of the DRASTIC vulnerability mapping methodology, the fuzzy hierarchy methodology for contamination source mapping, and the computational simulation of the nitrate contamination risk map. The modification implemented in this study with the application of the widespread intrinsic vulnerability model (DRASTIC) is based on a series of innovations that have taken place on this model since the beginning of its use in groundwater vulnerability studies. In my doctoral research, I have applied a Modified simple overlaying approach to creating a land-use layer attempted of integrating it with the modified reliable DRASTIC framework to be the parameter number eight in the overlaying approach of this study. The updated composite DRASTIC-land use model was then able to determine specific groundwater vulnerabilities by including the introduced land use parameter. But in this study due to the difficulties of recognizing all possible sources of contaminants, the susceptibility is more directed towards a particular contaminant which is in this dissertation nitrate. Therefore, the land use layer has been created to represent possible nitrate contamination (PNC) in the study area.

Lastly, the chapter presents the following key findings on the level of groundwater studies in Jordan:

1. Improved the quality of input datasets through the reduction of errors.
2. Showed that spatially explicit methods can improve the analysis of pollution sources and risks in the study area and Jordan. As well as the SCD representation and the statistical methods for assessment and representation and analyzing groundwater quality data can determine the extent of pollution in a cost-effective manner.
3. Provided a comprehensive analysis of the potential application of the DRASTIC model in arid areas.
4. The other issue highlighted in this research is the suitability of the groundwater vulnerability model (DRASTIC) used in this study to map pollution susceptibility in an arid area, while the map used in Jordan does not match the nature of the contamination susceptibility, since the mathematical model (COP) used is suitable for wet areas and is not suitable for dry areas.

Recommendations

Groundwater vulnerability and contamination risk mapping are exemplary decision-making devices to be taken into account in land use planning and global water resource management programs. It is, therefore, necessary to follow plans for the creation of contamination susceptibility maps using an appropriate model that is suited to the geological, hydrogeological, and weather conditions of the study area. Through the study, it was found that the modified (DRASTIC-model) is the most suitable for the arid area and can simulate the sensitivity of the area to pollution, while (COP-model) failed to do so, due to its heavy dependence on the rate of precipitation to estimate the intrinsic groundwater vulnerability indexes. Updating, validating, studying, and analyzing geological, hydrogeological, and water quality data through scientific and accurate methods, and avoiding data errors that may lead to unreal results (for example the hydraulic conductivity).

Conclusion

The study demonstrates that intensive land use in arid areas imposes tremendous pressure on groundwater. The comprehensive investigations of the contamination loads, the potential protectiveness of the aquifers, and the anthropogenic activities indicate that the contamination load is primarily attributable to agricultural activities and accompanied by high susceptibility to contamination. This study shows that remote sensing and GIS techniques are powerful tools in groundwater studies, providing possibilities for the use of vast spatial data, especially in overlaying modeling techniques. And concluded that a significant proportion of the case study area is hazardous to contaminants, demonstrated by the vulnerability and nitrate contamination RISK Maps.

Groundwater vulnerability models are the most useful tools to simulate the various control factors that govern the surface contamination leaching process towards the aquifers. A study illustrating the significance of vulnerability models in dry areas, in the case of Jordan, and its contribution to groundwater sustainability. Two DRASTIC approaches (ordinary and agricultural DRASTIC) were performed with two scenarios. Sensitivity tests were applied to modify and examine the original theoretical weights and avoid subjectivity in ratings and ranges of the parameters. Real parameters weights were calculated for the two scenarios in each DRASTIC approach by different methodologies relying on the (GIS) and using the extracted random points with the values of the seven rated DRASTIC parameters-maps. Long average and maximum concentrations of nitrate, sulfate, and salinity were used to assess the DRASTIC results since agriculture is the main source of pollution in the area. The comparison between the COP model and the DRASTIC model indicates the appropriate use of DRASTIC in arid areas. The procedure, successfully applied in this study with a reasonably good match between the RISK map and the nitrate distribution in groundwater, appears to be accurate and has with the contamination load investigations a large potential to be applied worldwide.

References

Albinet, M. & Margat, J. (1970). Cartographie de la vulnerabilité à la pollution des nappes d'eau souterraine [*Mapping vulnerability to groundwater pollution*], *Bull. BRGM 2nd Ser. (2ème série)*, Vol.3, No.4, pp 13–22.

Aller, L., Bennet, T., Lehr, J. H., Petty, R. J. and Hackett, G. (1987). DRASTIC: A standardized system for evaluating groundwater pollution potential using hydrogeologic settings, *Doc. EPA/600/2-87/035, p.622, United States Environmental Protection Agency (USEPA).*

Babiker, I. S., Mohamed, M. A. A., Hiyama, T. & Kato, K. (2005). A GIS-based DRASTIC model for assessing aquifer vulnerability in Kakamigahara Heights, Gifu Prefecture, central Japan, *Science of the Total Environment*, Vol. 345, No. 1–3, pp. 127–40.

Bachmat, Y. & Collin, M. (1987). Mapping assess groundwater vulnerability to pollution. In Vulnerability of soil and groundwater to pollution. In W. van. Duijvenbooden & H.G. van. Waegeningh (eds.), *The Hague, Proceeding and Information No.30*, TNO Committee on Hydrological Research, The Hague, Netherlands, pp. 297–307.

Ball, D., MacDonald, A., Dochartaigh, B., Rio, M. del, Fitzsimons, V., Auton, C. & Lilly, A. (2004). WFD28 Development of a groundwater vulnerability screening methodology for the water framework directive 2004 [Final Report], *Project WFD28, SNIFFER,* Registered Office: Caledonian Exchange, 19a Canning Street, Edinburgh, EH3 8HE, (Accessed 22/6/2019), www. sepa.org.uk/pdf/groundwater/tools/vulnerabi lity_rgeport.pdf.

Balon, M. & Dehnad, F. (2006). Water crisis in arid and semi-arid regions- an international challenge, *Symposium in Tehran Iran*, No. 49, Tehran, Iran, pp. 1–13, (Accessed 12/2/2019), http://www.water-asar.de/index_htm_files/water_crisis.pdf.

Barber, C., Bates, L., Barron, R. and Allison, H. (1993). Assessment of the relative vulnerability of groundwater to pollution: a review and background paper for the conference workshop on vulnerability assessment, *J. Australian Geology and Geophysics*, Vol. 14, Nos. 2–3, pp.1147–1154.

BGR & MWI (2018). Explanatory notes for the groundwater vulnerability map of Jordan (middle and shallow aquifer systems), 'Technical Report no.14', *Project: Improved Groundwater Resources Management in Response to the Syrian Refugee Crisis (I-GWRM), BGR-no (05-2383)*, Amman, Jordan.

Box, G. E. P., & Draper, N. R. (1987). .*Empirical Model-Building and Response Surfaces, (Wiley Series in Probability and Mathematical Statistics)*, p.688. ISBN-10 : 0471810339; ISBN-13: 978-0471810339, John Wiley & Sons. Washington, DC.

Breiman, L. (2001). Statistical modeling: the two cultures, *Statistical Science*, Vol. 16, No. 3, pp.199–231.

Brundtland, H. (1987). *Report of the World Commission on Environment and Development: Our Common Future*, Hauff, V. (Ed.), UN through the Oxford University Press, Oslo.

Canter, Larry.W. (1987). Nitrates in groundwater from agricultural practice -causes, prevention, and clean–up. *Report to the United Nation Development Program (UNDP), University of Oklahoma, Norman, Oklahoma, USA.*

Canter, Larry.W. (1997). *Nitrates in Groundwater*, 1[st]ed., pp. 263, Publisher: Routledge & CRC Press, Boca Raton.

Carsel, R. F., Smith, C. N., Mulkey, L. A., Dean, J. D. & Jowise, P. (1984). *User's Manual for the Pesticide Root Zone Model (PRZM),* Release I. (EPA/600/3-84/109), U.S. Environmental Research Laboratory (USERL), Athens, Georgia.

Chen, J., Zhang, H., Qian, H., Wu, J. and Zhang, X. (2013). Selecting proper method for groundwater interpolation based on spatial correlation, *Fourth International Conference on Digital Manufacturing & Automation*, pp.1192–1195, Qingdao, China, doi: 10.1109/ICDMA.2013.282.

Civita, M. (1990). *La valutazione della vulnerabilitá degli acquiferi all' inquinamento.[Assessment of aquifer vulnerability to contamination].* Proceedings of 1st national Congress on *(*Protection and Management of Groundwater*), "Protezione e Gestione delle Acque Sotterranee: Methodologie Techologie e Obiettivi,"* Marano sul Panaro, Italy, Vol. 3, pp. 39-86.

Civita, M. & Maio De, M. (2004). Assessing and mapping groundwater vulnerability to contamination: The Italian "combined" approach, *Geofísica Internacional*, Vol. 43, No. 4, pp. 213–532.

Cohen, D. B., Fisher, C. & Reid, M. L. (1986). Groundwater contamination by toxic sub stances: A California assessment, *In Evaluation of Pesticides in Groundwater,* Garner, W. Y., Honeycutt, R. C. and Nigg, H. N., eds. ACS Symp. Series 315, pp. 499-529, American Chemical Society (ACS), Washington, D.C.

Daniel J., G., Lisa A., S., Ali, S. & Ayman, J. (2013). *Groundwater-Level Trends and Forecasts, and Salinity Trends, in the Azraq, Dead Sea, Hammad, Jordan Side Valleys, Yarmouk, and Zarqa Groundwater Basins, Jordan*, Open-File., USGS, U.S. Department of the Interior, Prepared in cooperation with the U.S. Agency for International Development and the U.S. Army Corps of Engineers, Virginia.

Dawoud, M. A.; Sallam, O. M. (2012). sustainable groundwater resources management in arid regions: Abu Dhabi Case Study; Environmental Agency Abu Dhabi Report; Environmental Agency Abu Dhabi: Abu Dhabi,United Arab Emirates, *Presented in Ajman international Environmental Conference (Sustainable Development and Green Environment)*, 30-31Jan, (Accessed 30/oct/2019) from (https://www.researchgate.net/publication/260637853).

Dean, J. D., Huyakorn, P. S., Donigian, A. S. J., Voss, K. A., Schanz, R. W., Meeks, Y. J. & Carsel, R. F. (1989). Risk of unsaturated/saturated transport and transformation of chemical concentration (RUSTIC), Vol. I and II, (EPA/600/3-89/048a), *U.S. Environmental Protection Agency (USEPA), contract no.(68033604), Athens (30613), Georgia*, (Accessed 2/5/2019), https://nepis.epa.gov.

Denny, S. C., Allen, D. M. & Journeay, J. M. (2007). DRASTIC-Fm: A modified vulnerability mapping method for structurally controlled aquifers in the southern Gulf Islands, British Columbia, Canada, *Hydrogeol. Journal,* Vul. 15 No.3, pp. 483–493.

Doerfliger, N., Jeannin, P. Y. & Zwahlehn, F. (1999). Water vulnerability assessment in karst environments: A new method of defining protection areas using a multi-attribute approach and GIS tools (EPIK method), *Environ. Geol.,* Vol.39, No.2, pp 165–176.

Ducci, D. (2018). An easy-To-use method for assessing nitrate contamination susceptibility in groundwater, *Geofluids Journal-Hindawi journals*, Vol. 2018, No. 1, Article ID 1371825, pp12, https://doi.org/10.1155/2018/1371825.

Enfield, C. G., Carsel, R. F., Cohen, S. Z., Phan, T. & Walters, D. M. (1982). Approximating pollutant transport to ground water, *Ground Water*, Vol. 20, No. 6, pp. 711–722.

Evans, Thomas. Anders. B.S.M.E.M.S.E. (1995). *A Spatial and Statistical Assessment of The Vulnerability of Texas Groundwater To Nitrate Contamination*, PhD Dissertation, supported by the United States Environmental Protection Agency (USEPA), the Principal Investigator Prof. David R. Maidment, Center For Research in Water Resources, Bureau of Engineering Research, The University of Texas at Austin, J. J. Pickle Research Campus. Austin, TX 78712-4497. (Accessed 1/2/1019), http://civil.ce.utexas.edu/centers/crwr/reports/online.html.

Faybishenko, B., Nicholson, T., Shestopalov, V., Bohuslavsky, A. and Bublias, V. (2013). Groundwater vulnerability, Chernobyl nuclear disaster, *Journal of Chemical Information and Modeling*, Vol. 53, No. 9, pp. 1689-1699, co-publication between the American Geophysical Union (AGU) and John Wiley & Sons, Inc., https://doi. 10.1017/CBO9781107415324.004.

Foster, S. (2001). *Assessment of Groundwater Pollution Risk Contaminant Load*, pp. 13, Lima/Peru, (Accessed 1/3/2020), http://citeseerx.ist.psu.edu/index.

Foster, S., Chilton, J., Nijsten, G. & Richts, A. (2013). Groundwater: a global focus on the "local resource," *Current Opinion in Environmental Sustainability*, Vol. 5, No. 6, pp. 685–695, Doi:10.1016/j.cosust.2013.10.010.

Foster, S., Hirata, R., (1988). *Groundwater Risk Assessment: A Methodology Based on Available Data, Final Report* pp 86, World Health Organisation (WHO) & Pan American Health Organization (PAHO). WHO–PAHO, Lima, Peru, (Accessed 24/6/2020), https://scholar.google.com/citations?user=drEeWFcAAAAJ&hl=en&oi=sra.

Foster, S., Hirata, R., Gomes, D., D'Elia, M., Paris, M., (2002). *Groundwater Quality Protection: A guide for Water Utilities, Municipal Authorities, and Environment Agencies, Groundwater Quality Protection*. pp.114, ISBN: 0821349511, The International Bank for Reconstruction and Development/The World Bank, Washington DC 20433. https://doi.org/10.1596/0-8213-4951-1.

Foster, S. S. D. (1987). Fundamental concepts in aquifer vulnerability, pollution risk and protection strategy, W. van. Duijvenbooden & H. G. van. Waegeningh (eds.), *Vulnerability of Soil and Groundwater to Pollution*, TNO Committee on Hydrological Research, The Hague, Proceeding and Information No. 38, Netherlands, pp. 69–86.

Freeze, R. A. and Cherry, J. A. (1979). *Groundwater*, Brenn, C. and McNeily, K. (Eds.), Prentice-Hall, Inc., Englewood Cliffs, NJ 07632, New Jersey.

Garrett, P., Williams, J., Rossoll, C. & Tolman, A. (1989). Are ground water vulnerability classification systems workable? In: *National Ground Water Association Columbus Proceedings of the FOCUS Conference on Eastern Regional Ground-Water Issues,* Kitchener, Ontario, Canada, pp 329–343.

Gazal, O. (2015). *Groundwater Contamination of Shallow Aquifer Due to Irrigation Activities in Dhuleil and Khalidiyyeh Area in Jordan*, (supervisor: Prof. M. Shatanawi and Advisor: Prof. N. Lamaddalena) [Master thesis], issue number. 759, pp. 160, Istituto Agronomico Mediterraneo di Bari Collection/International Centre for Advanced Mediterranean Agronomic Studies, Valenzano. Bari. Italy.

Gazal, O. (2021a). Groundwater deterioration in arid agricultural area under the pressure of climate change and high refugee inflow. Case study: Jordan, *International Journal of Hydrology Science and Technology*, Vol.12, No.3, pp. 253-315, DOI: https://dx.doi.org/10.1504/IJHST.2021.117547.

Gazal, O. (2021b). *Comprehensive Risk Assessment for Sustainable Groundwater Management in Arid Areas, Case Study: Agricultural Eastern Jordan desert*, PhD Dissertation. Discipline: Environmental Sciences, with specialization in: Hydro-geological and Environmental Applied Engineering, (Supervisor: Dr. Sandor Szalai and co-supervisor: Dr. Tamás Szegi), pp. 140, Hungarian University of Agriculture and life sciences (formerly Szent Istvan University), Godollo. Hungary.

Gazal, O. & Eslamian S. (2021a). Comprehensive groundwater risk assessment case study: Arid Northern Jordan agricultural areas. *International Journal of Hydrology Science and Technology (IJHST).* Vol.12, No.4, pp. 382–447, doi: https://dx.doi.org/10.1504/IJHST.2021.118319 (Received:07Sep2020.

Gazal, O. & Eslamian, S. (2021b). Assessing groundwater vulnerability based on modified DRASTIC-model and COP-model in arid areas. *International Journal of Hydrology Science and Technology (IJHST).* Vol.13, No.2, pp 146-190, doi: 10.1504/IJHST.2021.10035061.

Gogu, R. C. & Dassargues, A. (2000). Current and future trends in groundwater vulnerability assessment, *Environmental Geology*, Vol. 39, No. 6, pp. 549–559.

Goldberg, V. M. (1983). Natural and technogenic factors of groundwater protectability. *Moscow Soc. of Nature Investigators Bull. No. 2,* pp. 103–110.

Goldberg, V. M. (1987). *Interrelation of Groundwater Contamination and Natural Environment, Gidrometeoizdat, Leningrad* (in Russian).

Goldscheider, N. (2005). Karst groundwater vulnerability mapping: Application of a new method in the Swabian Alb, *Germany, Hydrogeol. J.*, Vol. 13, No. 4, pp. 555–564.

Gowler, A. (1983). Underground purification capacity. In: *Groundwater in Water Resources Planning, Proceedings of the Koblenz Symposium. IAHS Publication 142, Vol. 2,* International Association of Hydrological Sciences, Wallingford, UK, pp.1063-1072.

HC & USEPA (2012). *Draft Final Report Identification and Evaluation of Existing Models for Estimating Environmental Pesticide Transport to Groundwater,* Health Canada (HC) and US Environmental Protection Agency (USEPA), (Accesed 2020 Feb 18). http://www.epa.gov/sites/prod.

Hobbler, M., Margane, A., Almomani, M., Subah, A. (2001). *Groundwater Resources of Northern Jordan, Contributions to the Hydrology of Northern Jordan.* By the

Technical Cooperation Project between the BGR (Federal Institute for Geosciences and Natural Resources_Hannover) and WAJ (Water Authority of Jordan), (Project number 89.2105.8), Vol.4, pp. 174, Amman, Jordan.

Hölting, B., Haertle, T., Hohberger, K. H., Nachtigall, K. H., Villinger, E., Weinzierl, W. & Wrobel, J. P. (1995). *Konzept zur Ermittlung der Schutzfunktion der Grundwasserüberdeckung. Geologisches Jahrbuch Reihe C, Band C63 [Concept for determining the protective function of groundwater cover. Geological Yearbook Series C, Volume C63].* Schweizerbart Science Publishers, Stuttgart, Germany.

Hoyer, B. E. & Hallberg, G. R. (1991). *Groundwater Vulnerability Regions of Iowa, Special Map 11.* Iowa City: Iowa Department of Natural Resources, (Accessed 12/9/2020), http://publications.iowa.gov/25744/.

Hussain, M. H., (2004). *Assessment of Groundwater Vulnerability in an Alluvial Interfluve using GIS* [PhD thesis], India Institute of Technology Roorkee Roorkee-247-667, pp,247, India.

Jordanian Institute of Standards and Metrology (JISM) (2008). *Drinking Water Standards No. 286/2008*, 5th ed., Department of Standards and Specifications, PMJ, Prime Minister of the Hashemite Kingdom of Jordan, Amman, Jordan.

Josopait, V., Schwerdtfeger, B., (1979). Geowissenschaftliche Karte des Naturraumpotentials von Niedersachsen und Bremen: [mit Höhenlinien] 1:200000, Bremerhaven Grundwasser: Tiefliegende Rohstoffe. Erze, Steinkohle, Industriemetalle, CC.3110 [*Geoscientific map of the natural space potential of Lower Saxony and Bremen: [with contour lines] 1:200000, Bremerhaven Groundwater: Low-lying raw materials. Ores, hard coal, industrial metals, CC.3110*], Publisher: Niedersächs. Landesamt f. Bodenforschung, Hanover. Germany.

Jury, W. A. & Roth, K. (1992). *Transfer Functions and Solute Movement through Soil: Theory and Applications,* Birkhaeuser Verlag. Berlin 1990; 226 Seiten, DM 65,-. ISBN 3-7643-2509-7 https://doi.org/10.1002/jpln.19921550218.

Jury, W. A., Spencer, W. F. & Farmer, W. J. (1983). Behavior assessment model for trace organics in soil: I. Model description, *Journal of Environmental Quality*, Vol. 12, No.4, pp. 558–564, (Accessed 2/11/2020) https://www.ars.usda.gov/arsuserfiles/20361500/pdf_pubs/P860.pdf.

Kazemi, G. A., Lehr, J. H. & Perrochet, P. (2005). *Groundwater Age*, pp 325, ISBN-13: 978-0-471-71819-2, John Wiley & Sons, INC, printed in the United States of America.

Kissel, D. E., Bidwell, O. W., Kientz, J. F., (1982). Leaching classes in Kansas soil. Bulletin No. 641, pp. 14, Contribution number: 82-416-B, *Kansas State University Agricultural Experiment Station and Cooperative Extension Service*, Kansas State University, Kansas, USA, (Accessed date 2/5/2020), https://www.ksre.k-state.edu.

Lee, S., Lee, D., Choi, S. H., Kim, W. Y. & Lee, S. G. (1998). Regional groundwater pollution susceptibility analysis using DRASTIC system and lineament density, *18th Annual ESRI International User Conference,* San Diego, California (USA), (Accessed 2/8/2020), https://proceedings.esri.com/library/userconf/proc98/proceed/to200/pap171/p171.htm.

Leonardo, R. A., Knisel, W. G. & Still, D. A. (1987). GlLEAMS: Groundwater loading effects of agricultural management systems, *American Soc. Agric. Eng.*, vol. 30, No.

5, pp. 1403–18. (Downloaded from: https://elibrary.asabe.org/abstract.asp?aid=30578).

Loague, K. R. H., Abrams, S. N., Davis, A., Nguyen & Stewart, I. T. (1998). A case study simulation of DBCP groundwater contamination in Fresno County, California 1. Leaching through the unsaturated subsurface, *Journal of Contaminant Hydrology*, Vol. 29, No. 2, pp. 109–36.

Lodwick, W. A., Monson, W. & Svoboda, L. (1990). Attribute error and sensitivity analysis of map operations in geographical informations systems: Suitability analysis, *International Journal of Geographical Information Systems*, Vol. 4, No. 4, pp. 413–28.

Margane, A. (2003). Management, protection, and sustainable use of groundwater and soil resources in the Arab region: Guideline for groundwater vulnerability mapping and risk assessment for the susceptibility of groundwater resources to contamination, management, vol. 4, BGR, and ACSAD, *Damascus*, pp. 242.

Margane, A., Hobbler, M., Almomani, M. and Subah, A. (2002). *Contributions to the Hydrogeology of Northern and Central Jordan.* Hydrogeologie, Ingenieurgeologie, Hannover, Stuttgart, Germany, Geologisches Jahrbuch Reihe C, Band C, Vol. 68, 52pp, Schweizerbart Science Publishers, Germany, ISBN: 978-3-510-95890-0 [online] http://www.schweizerbart.de/publications/detail/artno/186046800.

Margane, A., Hobler, M. and Subah, A. (1997). *Special Report No. 3: Mapping of Groundwater Vulnerability and Hazards to Groundwater in the Irbid Area*, Report prepared by WAJ & BGR, Technical Cooperation Project 'Advisory Services to the Water Authority of Jordan', No. 3, p.50.

Margane, A., Hobler, M. and Subah, A. (1999). Mapping of groundwater vulnerability and hazards to groundwater in the Irbid area, N Jordan, *Zeitschrift Angewandte Geologie.*, Vol. 45, No. 4, pp.175–187.

Margat, J. (1968). Vulnerabilite des nappes d'eau souterraine a la pollution *[Ground water vulnerability to contamination]. Bases de la cartographic, (Doc.) BRGM,* 68 SGL 198 HYD, Orleans, France [in French].

Margat, J., Foster, S. & Droubi, A. (2006). Concept and importance of non-renewable resources, in D.P. Foster, S. & Loucks (ed.), *A Guidebook on Socially-Sustainable Management for Water-Policy Makers*, (UNESCO) United Nations Educational, Scientific and Cultural Organization, Place de Fontenoy, 75352 Paris 07 SP, France, Composed by Marina Rubio, 93200 Saint-Denis, Vol. 10, pp. 13–24, (accessed 1/2/2020), https://unesdoc.unesco.org.

McLay, C., Dragten, R., Sparling, G. & Selvarajah, N. (2001). Predicting groundwater nitrate concentrations in a region of mixed agricultural land use: a comparison of three approaches. *Environmental Pollution*, Vol. 115, No. 2, pp. 191–204.

McLean, J. E., Sims, R. C., Doucette, W. J., Caupp, C. R. & Girenney, W. J. (1988). Evolution of mobility of pesticides in soil using U.S. EPA methodology, *ASCE. Journal of Environmental Engineering*, Vol. 114, No. 3, pp. 689–703.

Melloul, A. J. & Collin, M. (1998). Proposed index for aquifer water quality assessment: the case of Israel'Sharon region, *Journal of Environmental Management*, Vol. 54, No.2, pp. 131–42.

Merchant, J. W. (1994). GIS-based groundwater pollution hazard assessment: a critical review of the DRASTIC model, *Photogramm. Eng. Remote. Sens*, Vol. 60, No. 9, pp. 1117–1127.

Ministry of Energy and Mineral Resources (MEMR) (2017). *National Mapping Project (Geological Maps Index, Updated Dec. 2017)*, Amman, Jordan.

Ministry of Planning and International Cooperation (MOPIC) (2014). *Needs Assessment Review of the Impact of the Syrian Crisis on Jordan, Host Community Support Platform (HCSP) Report*, 122pp, United Nations.

MOA (1994). *The Soils of Jordan. Report of the National Soil Map and Land Use Project.* Undertaken by the Ministry of Agriculture (MOA), Huntings Technical Services Ltd., and European Commission. Level One, Level Two, Level Three, and JOSCIS Manual, Ministry of Agriculture, Amman, Jordan.

MOA (2020). *Agricultural Intensity Maps*, GIS Unit, Department of Rangeland, Ministry of Agriculture (MWI), Amman, Jordan.

Moore, J. S. (1988). Seeppage: A system for early evaluation of pollution potential of agricultural ground water environments. *Engeneering Geology Technical Note 5 (Revision1).* Washington, D.C.: U.S. Department of Agriculture, Soil Conservation Service.

Morris, B. & Foster, S. (2001). *Assessment Of Groundwater Pollution Risk Contaminant Load*, published report, pp.18, Lima/Peru. (Online source accessed date 10/8/2020), http://citeseerx.ist.psu.edu/viewdoc/download?doi=10.1.1.198.5322&rep=rep1&type =pdf.

MWI (2013). Feasibility study and ESIA for Zarqa Governorate water wells rehabilitation, *Mostaqbal Engineering and Environmental Consultants* vol. 1, Amman, Jordan.

MWI and JWA (2010). Feasibility study and ESIA for Zarqa Governorate water wells rehabilitation. *Mostaqbal Engineering and Environmental Consultants*. Groundwater Model Report, Vol. 2, No. 2, Amman, Jordan.

MWI/USAID/ARD (2001). Hydrogeological impacts of overpumping and assessment of groundwater management options in the Amman-Zarqa Highlands, *Water Resource Policy Report: Groundwater Management Component*, Amman, Jordan.

Napolitano, P. (1995). *GIS Aquifer Vulnerability Assessment in the Piana Campana, Southern Italy, using the DRASTIC and SINTACS Methods* [M.Sc. thesis], International Institute for Aerospace Survey and Earth Sciences, Netherlands'.

Napolitano, P., Fabbri, A. G. (1996). Single parameter sensitivity analysis for aquifer vulnerability assessment using DRASTIC and SINTACS. In *Proceedings of the Second HydroGIS Conference Titled: Application of Geographical Information Systems in Hydrology and Water Resources Management (Proceedings, Vienna Conference),* Kovar, K., Nachtnebel, H.P., Eds.; IAHS Publication no. 235; IAHS: Wallingford, UK, pp. 559–566.

NASA (2019). *ASTER Global Digital Elevation Model V003*. Distributed by NASA EOSDIS Land Processes DAAC, https://doi.org/10.5067/ASTER/ASTGTM.003. Accessed 2022-07-08.

National Research Council (NRC) (1993). *Ground Water Vulnerability Assessment: Predicting Relative Contamination Potential Under Conditions of Uncertainty*,

Ground Water Vulnerability Assessment, The National Academies Press, Washington, DC.

National Research Council (NRC) (1994). *Ground Water Recharge Using Waters of Impaired Quality, Ground Water Recharge Using Waters of Impaired Quality*, Washington, D.C.

National Water-Quality Assessment Program (NAWQA) (1999). Improvements to the DRASTIC groundwater vulnerability mapping method, *U.S. Department of the Interior, (USGS). Fact sheet FS-066-99.* https://doi.org/10.3133/fs06699.

Navulur, K. C. S. & Engel, B. A. (1994). GIS in statewide ground-water vulnerability evaluation to pollution potential, *Seminar Publication: National Conference on Environmental Problem-Solving with Geographic Information Systems,* (EPA/625/R-95/004), United States Environmental Protection Agency (US-EPA), Cincinnati, Ohio, pp. 66–74.

Navulur, K. C. S. & Engel, B. A. (2003). Predicting spatial distributions of vulnerability of Indiana State aquifer systems to nitrate leaching using a GIS, *National Center for Geographic Information and Analysis (NCGIA) at its three consortium sites, UCSB; the State University of New York, Buffalo; and the University of Maine.*, National Service Center for Environmental Publications (NSCEP), (Accessed 22 February 2020), http://www.ncgia.ucsb.edu/SANTA_FE_CD-ROM/sf_papers/navulur_kumar/my_paper.html.

Nguyet, V. T. M. & Goldscheider, N. (2006). A simplified methodology for mapping groundwater vulnerability and contamination risk, and its first application in a tropical karst area, Vietnam, *Hydrogeology Journal*, Vol. 14, No. 8, pp. 1666–75.

Nobre, R. C. M., Rotunno Filho, O. C., Mansur, W. J., Nobre, M. M. M. and Cosenza, C. A. N. (2007). Groundwater vulnerability and risk mapping using GIS, modeling and a fuzzy logic tool, *Journal of Contaminant Hydrology*, Vol. 94, Nos. 3–4, pp.277–292.

Ouedraogo, I. Defourny, P. & Vanclooster, M. (2016). Mapping the groundwater vulnerability for pollution at the pan African scale, *Science of the Total Environment*, Vol. 544, No.11, pp. 939–53.

Pettyjohn, W. A., Savoca, M. & Self, D. (1991). Regional assessment of aquifer vulnerability and sensitivity in the conterminous United States. *Report EPA-600/S2-91/043*. Ada, Oklahoma: U.S. Environmental Protection Agency (EPA).

Price, M. (2013). *Introducing Groundwater*, publication of *Chapman & Hall, 2-6 Boundary Row, London SEI8HN, UK*, Springer-Science+Business Media, B. Y., London. ISBN: 9780412485008, Doi:10.4324/9780203761045.

Ramolino, L. (1988). 'Development o f a Nitrate Pollution Index for Ground Water', Master thesis, University o f Oklahoma, Norman, Oklahoma.

Ramolino, L. & Canter, L. W. (1990). *Aquifer Pollution Vulnerability Mapping*. The Environmental and Ground Water Institute, University of Oklahoma, Norman, Oklahoma.

Ramos, J. A. & Castillo, R. R. (2003). Aquifer vulnerability mapping in the Turbio river valley, Mexico: A validation study, *Geofisica International*, Vol. 42, No. 1, pp. 141–56.

Ramos-Leal, J. A. & Rodríguez-Castillo, R. (2003). Aquifer vulnerability mapping in the Turbio river valley, Mexico: A validation study, *Geofísica Internacional*, Vol. 42, No. 1, pp. 141–56.

Rao, P. S. C., Hornsby, A. G. & Jessup, R. E. (1985). Indices for ranking the potential for pesticide contamination of groundwater. *Soil Crop Science Society Florida Proceedings* Vol. 44,pp. 1-8.

RJGC (2020). *Administrative Divisions*, RJGC Royal Jordanian Geographic Centre [online] http://rjgc.gov.jo/rjgc_site/%0D (accessed 2 June 2020).

Rosen, L. A. (1994). Study of the DRASTIC methodology with emphasis on Swedish conditions, *Ground Water*, Vol. 32, No. 2, pp. 278 –85.

Rubel, F., Brugger, K., Haslinger, K. and Auer, I. (2017) The climate of the European alps: shift of very high resolution Köppen-Geiger climate zones 1800–2100, *Meteorologische Zeitschrift,* Vol. 26, No. 2, pp.115–125.

Sacha, L., Fleming, D. & Wysocki, H. (1987). Survey of pesticides in selected areas having vulnerable ground water in Washington State. *EPA/91/9-87/169*. Seattle, Washington: U.S. Environmental Protection Agency, Region X.

Secunda, S., Collin, M. & Melloul, A. (1998). Groundwater vulnerability assessment using a composite model combining DRASTIC with extensive agricultural land use in Israel's Sharon region, *Journal of Environmental Management*, Vol. 54, pp. 39 – 57.

Sotornikova, R. & Vrba, J. (1987). Some remarks on the concept of vulnerability maps. In Vulnerability of soil and groundwater to pollution (W. van Duijvenbooden and H.G. van Waegeningh, eds.), *TNO Committee on Hydrological Research, The Hague, Proceeding and Information*, No. 38, pp.471-476.

Srivastava, R. (2013). Groundwater resources under a changing climate, *Journal of the Indian Institute of Science*, Vol. 93, No. 2, pp. 251–64.

Steenhuis, T. S., Pacenka, S. & Prter, K. S. (1987). MOUSE: A management model for evaluation groundwater contamination from diffuse surface sources aided by computer graphics, *Appl. Agr. Res*, Vol. 2, pp. 277–89.

Teso, R. R., Younglove, T., Peterson, M. R., Sheeks, D. L. & Gallavan, R. E. (1988). Soil taxonomy and surveys: classification of areal sensitivity to pesticide contamination of groundwater, *Journal of Soil and Water Conservation*, Vol. 43, No. 4, pp. 348–52.

Thaddeus, T. (2009). All models are right…most are useless, *Joint Statistical Meeting,* (Accessed 8 January 2020) https://statmodeling.stat.columbia.edu/wp-content/uploads/2012/03/tarpey.pdf.

U.S. Environmental Protection Agency (USEPA) (1993). *A Review of Methods for Assessing Aquifer Sensitivity and Ground Water Vulnerability to Pesticide Contamination (Draft),* Washington, D.C.

U.S. General Accounting Office (1991). *Groundwater Protection: Measurement of Relative Vulnerability of Pesticide Contamination.* Washington, DC, GAO/PEMD-92-8.

UNESCO-IHP (2004). *Groundwater Contamination Inventory: A METHODOLOGICAL GUIDE with a Model Legend for Groundwater Contamination Inventory and Risk Maps*, A. Zaporozec (ed.), IHP-VI, SE., International Hydrological Programme (IHP), UNESCO, Paric France.

USDA (2014). Keys to soil taxonomy, *Soil Conservation Service*, 12th ed., p.410 [online] http://www.nrcs.usda.gov/Internet/FSE_DOCUMENTS/nrcs142p2_051546.pdf (accessed 17August 2020).

Vías, J. M., Andreo, B., Perles, M. J., Carrasco, F., Vadillo, I., Jiménez, P. (2006). Proposed method for groundwater vulnerability mapping in carbonate (karstic) aquifers: the COP method. *Hydrogeology Journal*, Vol. 14, No. 6, pp. 912–925, https://doi.org/10.1007/s10040-006-0023-6.

Villumsen, A., Jacobsen, O. S. & Sonderskov, C. (1983). Mapping the vulnerability of groundwater reservoirs with regards to surface pollution. *Danmarks Geologiske Undersogelse, Arbog,* 17-38, Denemark, Copenhagen.

Vrana, M. (1968). Ochrana prostyc podzemnich vod v Cechach a na Morava.Vysvetlivsky k mape 1:500000 [*Protection of prostyc underground waters in Guilds and Moravia. Explanatory notes to map 1:500000*]. *Wat. Res. Plan Cent.,* Praha (in Ceco).

Vrana, M. (1984). Methodology for construction of groundwater protection maps, Lecture for UNESCO/UNEP Proj. PLCE3/29, Moscow, September 1981. *Hydrogeological Principles of Groundwater Protection,* Vol. 1, E. A. Kazlovsky Edit. and Chief, pp. 147–149, UNESCO/UNEP, Moscow.

Vrba, J. and Zaporožec, A. (1994) *Guidebook on Mapping Groundwater Vulnerability – IAH International Contributions to Hydrogeology*, 16th ed., Vol. 16, pp. 131, Heise Publication, Hannover/FRG, Heise, Germany.

Wagenet, R. J. & Huston, J. L. (1989). LEACHM: A finite-difference model for simulating water, salt and pesticide movement in the plant root zone, *Continuum Version 2,* (Ithaca, NY: New York State, Water Resources Institute, Cornell University, USA).

Wei, M. (1998). Evaluating AVI and DRASTIC for assessing pollution potential in the Lower Fraser Valley, British Columbia: Aquifer vulnerability and nitrate occurrence, *CWRA 51st Annual Conference Proceedings, Mountains to Sea: Human Interaction with the Hydrologic Cycle,* Victoria, BC. (Online accessed 20/1/2020), http://d.healthysocialnews.com/537.html.

WIS (2021). *Water Information System.* Amman Jordan: Ministry of Water and Irrigation, technical Affairs, The Hydrological and Meteorological Information Systems, Studies Directorate. (Retrieved from www.mwi.gov.jo).

Zaporozec, A. (Ed.) (1985). *Groundwater Protection Principles and Alternatives for Rock County, Wisconsin.* Madison, WI: Wisconsin Geological and Natural History Survey. (Accessed 1/9/2018) https://wgnhs.wisc.edu/catalog/series/educational-series.

Zaporozec, A. (2002). Groundwater contamination inventory: methotological guide. *UNESCO/IHP*, p. 160; (*IHP-VI, Series on Groundwater*, 2).

Zhou, J., Li, G., Liu, F., Wang, Y. & Guo, X. (2010). DRAV model and its application in assessing groundwater vulnerability in arid area: A case study of pore phreatic water in Tarim Basin, Xinjiang, Northwest China, *Environmental Earth Sciences*, Vol. 60, No. 5, pp. 1055–63.

Zwahlen, F. (2004). *COST Action 620 Vulnerability and Risk Mapping for the Protection of Carbonate (karst) Aquifers - Final Report.*, François. Zwahlen (ed.), ISBN: 92-894-6416-X, European Union (EU)/Directorate-General for Research, Belgium, (Accessed 22/5/2020), https://doi.org/PNR61.

Chapter 8

Tsunami Forecasting in Asia and the Middle East

Saeid Eslamian[1], Mousa Maleki[2*] and Nasir Ahmad Rather[3]

[1]Department of Water Engineering, College of Agriculture,
Isfahan University of Technology, Isfahan, Iran
[2]Civil, Architectural, and Environmental Engineering Department,
Illinois Institute of Technology, Chicago, Illinois, USA
[3]Department of Civil Engineering,
Baba Ghulam Shah Badshah University,
Rajouri, Jammu and Kashmir, India

Abstract

The rapid prediction of spatial distribution from the near-field tsunami is very important for tsunami risk characteristics. Despite significant improvements in the past decade, physics-based numerical models are still computational. Uncertainty is an important challenge in the risk analysis of tsunamis. Tsunami elevations are affected by complex factors and are continuously changing. Tsunami height has random features. This study suggests that water depth (related to sea topography) is the most important factor affecting the height of the tsunami. But around the world, a significant area of marine topography has not been measured. Therefore, you need to use uncertainty to consider the depth of water. Asia and the Middle East have both volcanic tsunamis and some popular, most economical, and rapidly developing beaches in the world. This helps examine the risk of tsunamis in Asia and the Middle East.

---

[*] Corresponding Author's Email: mmaleki1@hawk.iit.edu.

In: Handbook of Uncertainty in Eurasian Forecasting (HEF)
Editors: Saeid Eslamian, Mir Bintul Huda and Faezeh Eslamian
ISBN: 978-1-68507-966-6978

Keywords: fault, tsunami, prediction, Asia, Middle East

Introduction

Marine hazards forecasting and warning is the least effective way to deal with marine hazards. Marine hazards include physical and biological hazards. The physical and hydrodynamic properties of the environment play a major role even in how biological hazards such as algal blooms are constructed. Some marine hazards affect the marine environment and its shores, besides affecting miles on land. Climatic conditions on land such as hurricanes, droughts, and floods are all affected by ocean cycles and oceanic climate changes.

Tsunami forecasting and warning systems, not only reduce economic costs but also offer a range of economic and social development programs for coastal areas. Therefore, forecasting and warning of marine hazards play a key role in development plans. Ocean, marine biology, marine phenomena modeling, coastal engineering, ocean engineering, satellite oceanography, geological management, marine chemistry, and marine physics are the main specialties for providing prediction and warning of marine hazards. Many beaches within the area are susceptible to the tsunami. The tsunami is a Japanese word for harbor waves (Dawson et al., 1993).

A tsunami is a long wave whose period varies from a few minutes to a few hours. Tsunami wavelengths can range from a few hundred kilometers to several thousand kilometers. The height of a tsunami wave at sea and offshore is very low and is usually undetectable from a ship. Various phenomena can cause a tsunami wave, but the surrounding environment also plays an important part. Tsunamis do not form in shallow seas.

The movement of large faults in the seas and oceans and the creation of large earthquakes can cause tsunami waves. If the fault moves in such a way as to create a vertical height difference and the impact on the water in the bed is large enough, a tsunami wave is formed. Therefore, large earthquakes (magnitude of 7 and larger) usually cause tsunami waves.

Another cause of a tsunami is a large amount of sediment falling into the seabed or landslides. Just as landslides occur on steep mountain slopes, they can occur on steep seabeds and oceans. A tsunami wave occurs if a large volume of sediment is suddenly displaced, as shown in figure 1.

There are many islands in the oceans that are very high and have a small cross-section, if an island breaks and falls, it will create a very large tsunami wave. Occasionally gases collected in sedimentary layers erupt from the fracture site (volcanic eruption). Sudden eruptions with large volcanic eruptions can also cause tsunami waves. Seabed volcanoes are also capable of creating tsunami waves.

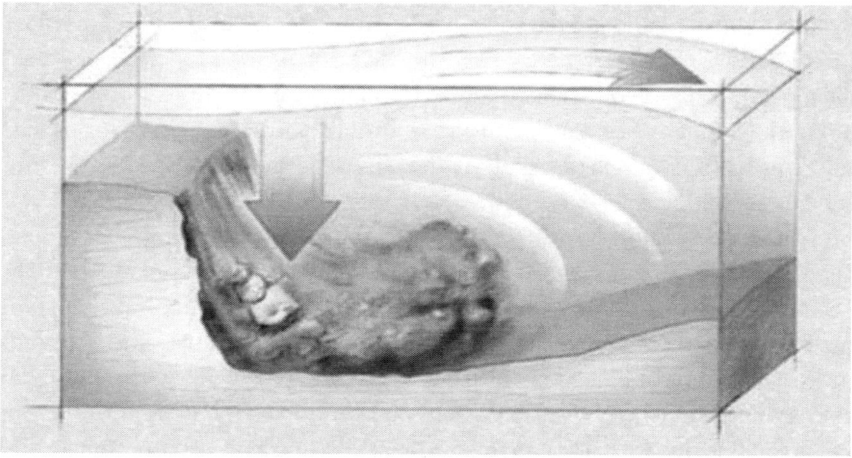

Figure 1. Process of tsunami caused by sudden sediment disposal.

Also, the impact of large meteorites on the seas and oceans can cause a tsunami wave. However, the two main causes of tsunamis in the seas and oceans are earthquakes and landslides. Because a tsunami is in many cases, an unpredictable event, setting up and developing a tsunami alert system is one of the most important tasks in educating the public about a tsunami hazard. Educating people to understand the dangers of tsunamis can greatly help prepare coastal communities to avoid these dangers. Creating escape routes and showing how to avoid these dangers can save lives. This is especially important in areas such as the Oman Sea, where there is little time for warning.

Doing high-altitude construction is the best way to avoid tsunami waves. In some countries, sea walls and structures are built to deal with tsunami waves. Nevertheless, these walls may not be effective against waves that are too high. Therefore, such structures should not be taken lightly. Even the construction of some of these structures may cause more damage during a tsunami because seawater will encounter these obstacles on return and the water level will remain high inland for longer.

First, keep in mind that the tsunami has nothing to do with the weather or the hours of the day. Tsunamis can occur in sunny or cloudy weather, calm or stormy sea conditions, and night or day. So, if you live on the shores of the Oman Sea, always be prepared for such an event.

East Asia (EAMS), including the East China Sea and the Yellow Sea with Korea, China, Japan, and Taiwan, is one of the world's largest marginal seas in the world. This is mainly connected through the Japanese Ryukyu Islands to the Pacific. In EAMS, coastal and environmental processes are influenced by tidal currents and topography has an important effect on deep-sea tide modification. Approximately 7% of the global tidal energy in EAMS disappears. The overall release of daily tides from the north to the south is along the Chinese Sea shelf, and the semi-daily tide has a constant phase in breaking the East China Sea shelf, with phases and slopes of land enhancement (Lee et al., 2020).

Tsunami warning systems are designed to predict tsunami waves before attacking coastal areas and causing severe damage. The prediction method usually consists of two stages, namely estimation of earthquake source mechanism and numerical simulation of tsunami waves. To accurately predict the wave height and the arrival time, it is essential to make an earthquake source model correctly. For near-field areas in which tsunami information is not available in real-time, an operational approach is usually approved to predict the tsunami based on the magnitude of an earthquake and the tsunami database predicted (Greenslade & Titov, 2008; Greenslade et al., 2011; Kamigaichi, 2009; Rudloff et al., 2009; Setiyono et al., 2017). Before the earthquake, uniform slip models are made for the prevalence of tsunami, according to a tsunami earthquake, the length of the tsunami, length of the rupture, width, and slip are determined based on the scaling relationship with the earthquake (Blaser et al., 2010; Mai & Beroza, 2000; Murotani et al., 2008, 2013; Strasser et al., 2010; Wells & Coppersmith, 1994). The tsunami waves are numerically simulated and stored. After the earthquake, tsunami warnings are issued based on the scenario with the highest magnitude of the earthquake. Advanced research on local tsunami warning techniques focuses on rapid earthquake estimation (Melgar & Bock, 2013; Melgar et al., 2015, 2016), earthquake rupture (AN & Meng, 2016), and forecast of tsunami waves using uniform slip models. In addition, uniform slip models are also used by researchers to evaluate local tsunami risks (Ren et al., 2017; Wu & Huang, 2009).

Methods of Tsunami Prediction

Tsunamis are some of the malicious and costly risks for coastal areas around the world. The tsunami close to the field, which is a tsunami produced close to the coastline, includes a high risk for coastal communities because the first waves can enter within minutes (National Council of National Research 2011). To reduce damages and create flexible coastal communities, it is important to put a quick predictive capacity for a tsunami redistribution along the coast.

Physics-based numerical simulation is currently the most accurate method for predicting tsunami distribution. Although significant progress has been made over the past decade (LeVeque et al., 2011; Lin et al., 2015; Popinet 2015; Shi et al., 2012), these numerical physics-based models are still valid. For example, the assessment of the probability of tsunami (PTHA) requires a tsunami estimate for many scenarios to accurately measure the risk and relevant uncertainty (Mori et al., 2018). However, due to the computational load associated with physics-based numerical simulation, a logical tree approach is commonly used with the number of scenarios based on historical earthquake characteristics.

The issue with the tree logic approach is that it relies on expert judgment. On the other hand, to perform many scenario simulations, several studies use the reinforcement factor method, which can quickly estimate the maximum height of the maximum drainage of simulated marine tsunamis (Davies et al., 2018; Glimsdal et al., 2019). However, the constraints of the boosting approach are that the selection of sea reference points is somewhat mental, and this approach still requires numerical simulation to obtain the marine tsunami domain. The computational load associated with physics-based numerical simulation especially for predicting a near-field tsunami is a major obstacle. For this reason, pre-calculated simulation databases are widely used. These databases can perform a quick prediction by choosing the best-coordinated simulation or by interpolation between simulations immediately after the source mechanism is known (Kamigaichi 2011; Mulia et al., 2018; Setiyono et al., 2017). A problem with the database approach is that there can be significant errors in real-world scenarios that are not available on selected bases.

Response Level Method (RSM)

The response level method (RSM) is an effective statistical approach to creating a betting relationship between a set of input and output variables of a system (Myers et al., 2016). When the RSM model is made, the output can be estimated quickly across the sequence of input spaces. However, because the high-dimensional input requires a large number - which is exclusively expensive - RSM has not been implemented to predict tsunami distribution. For example, a tsunamigenic earthquake (input in a RSM model) is usually shown by 9 faults parameters (Figure 2). A complete factorial design is one of the largest experiments tested (DOE) designs used to measure the response of any possible combination of independent variables. If we design the synthetic tsunami scenarios using a complete factorial method with 9 fault parameters, simulation of 19,683 (= 39) is required. In addition, if the input/output shows a large nonlinear relationship, a higher level of DOE may be required, which requires much more simulations.

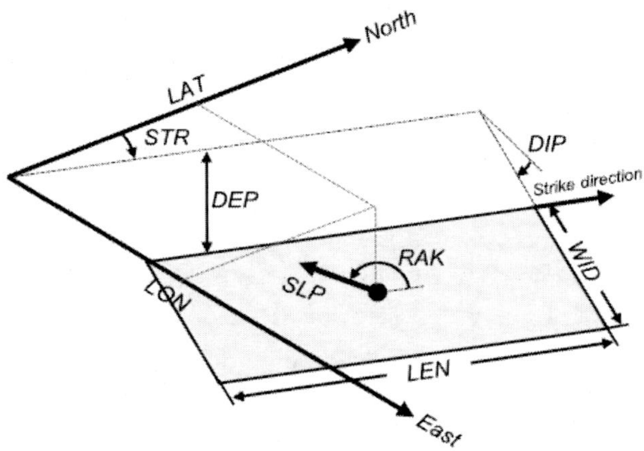

Figure 2. Schematic sketch of earthquake fault parameters: epicenter latitude (LAT), epicenter longitude (LON), fault length (LEN), fault width (WID), top-edge fault depth (DEP), strike angle (STR), dip angle (DIP), rake angle (RAK) and slip (SLP) (Lee et al., 2020).

Probabilistic Estimates of Tsunami Hazards

Conventionally, the risk of tsunami and risk analysis was performed using definitive methods. Due to the randomized tsunami nature, efforts have recently been made to develop probabilistic estimates of tsunami hazards, often called "probabilistic tsunami hazard assessment" (PTHA) (Geist and Parsons 2006; Thio et al., 2007; Gonzalez et al., 2013; GEIST and LYNETT 2014). This assessment is based on a long-implemented probabilistic risk assessment (PSA). PTHA methods are considered due to all possible tsunami parameters that may affect the formation of a tsunami. However, uncertainty associated with wave height modeling remains a key challenge in tsunami risk analysis (Zhang et al., 2018; Zhang and Cao 2015; Zhang and Lam 2014, 2015).

Currently, the assessment of the possible tsunami risk mainly uses maximum tsunami highlands, which occur near the beach or the depth of penetration in the earth (Gonzalez et al. 2013; Adams et al. 2014). Therefore, the maximum tsunami height is an essential parameter in the risk assessment of the tsunami. A good understanding of inherent uncertainty at the maximum tsunami height is important for assessing the risk of tsunami and risk assessment. The probabilistic method has been used in recent years to examine the maximum tsunami height (Burbidge et al., 2015). Mofjeld et al., (2007) studied tidal effects on the maximum height of the tsunami waves and found that the potential distribution of tsunami highlands depends on the primary tsunami domain and tidal properties.

Dong et al., (2017) examined the long-term tsunami height in Crescent City California. The severe tsunami heights, seen from 1938 to 2015, are assessed using six possible distributions (Gumbel distribution, Weibull distribution, maximum entropy distribution, distribution of lanemal, ultra-general value distribution, and generalized Pareto distribution). Using the Kolmogorov-Smirnov test (K-S), it was found that the value-added distribution was generally best suited for the maximum tsunami height.

Davies et al., (2015), McCloskey et al., (2007, 2008), Løvholt et al., (2012), and Goda et al., (2014) studied the non-uniform slip effect on the maximum tsunami height near the field and at the depth of the maximum collapse. Other factors affecting the wave field are also examined, such as Glimsdal et al., 2013, friction effects (Myers and Baptista 2001), and earth coverage roughness when a tsunami is presented on the coast (Kaiser et al., 2011). As mentioned in previous studies, various physical parameters have been investigated, which can affect the height of the tsunami, especially the source earthquake parameters.

One of the indirect but important factors that have not been studied on the maximum wave height that is systematically studied is the depth of water (bathymetry). This is only generally discussed in the literature (Burbidge et al., 2015; Geist 2009). The depth of water affects tsunami release and can determine the maximum tsunami height when the tsunami is spreading on the beach. However, almost all over the world, a significant area of marine topography has not been studied and measured. So, it is very necessary to use the uncertainty method to consider the depth of water, the most important factor affecting the maximum wave height.

It is important to determine the maximum tsunami height, as discussed above. In addition to the depth of the tsunami penetration and the speed of the tsunami flow, tsunami troops that affect the structure can also be calculated based on the maximum height of the tsunami wave (Yeh et al., 2014; Heintz and Robertson 2008). Therefore, the estimation of the maximum tsunami height is necessary to predict the amount of damage to essential structures such as nuclear power plants. There is a close relationship between the depth of water and tsunami height.

The Physical Process of Tsunami Propagation

The physical processes of the tsunami emission that occurs from the deep sea to shallow water are used to analyze the mechanism of maximum tsunami wave height. This will provide a physical basis for the maximum probability model of tsunami height. The body of water, which becomes a wave, is called the wave base. Under the base of the wave (where the depth of water is larger than half the wavelength), the body of the water is quiet without moving (Figure 3A).

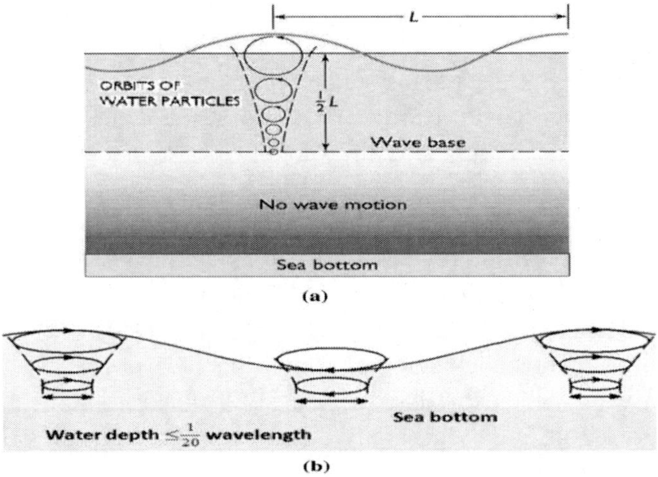

Figure 3. Deep-water waves and shallow-water waves (Zhang et al., 2020).

When the depth is larger than half the wavelength, the wave is deep water. The rate of diffusion of a deep-water wave is the same as a harmonic wave, and the speed of the wavelength is divided by the wave period,

$$u = \sqrt{gd} \tag{1}$$

where g is the gravitational acceleration. According to equation (1), the release rate is reduced as the depth of water is reduced.

Therefore, in the tsunami release process from the sea to land, the tsunami will change from a deep-water wave to a shallow water wave and the rate of release equals to $u = \sqrt{gd}$. From the salt water to the ground, the depth of water is continuously shallow, and its release rate decreases.

In the tsunami produced from the earthquake, the frequency and tsunami period only depends on the hypocenter, so the wave period in the deep wave of water is changing to the shallow wave of water. Due to the relationship "wave period = wave/wave velocity," when the speed decreases, the wavelength also lowers to maintain a constant period. The wave energy that is produced by the tsunami is almost constant. Therefore, the reduction of wave speed and wavelength causes a higher wave range. That is why the wave height goes to the beach as waves. As the wavelength is reduced, its range increases. When the wave range increases to a particular point, the wave breakpoint will cause interruption, and this interruption is the reason for common marine waves to be broken and turn into wave gaps. When the tsunami waves are broken, a water wall is produced by tens of meters in height due to high speed and large wave energy. The tsunami wave propagation process is shown from the sea in Figure 4.

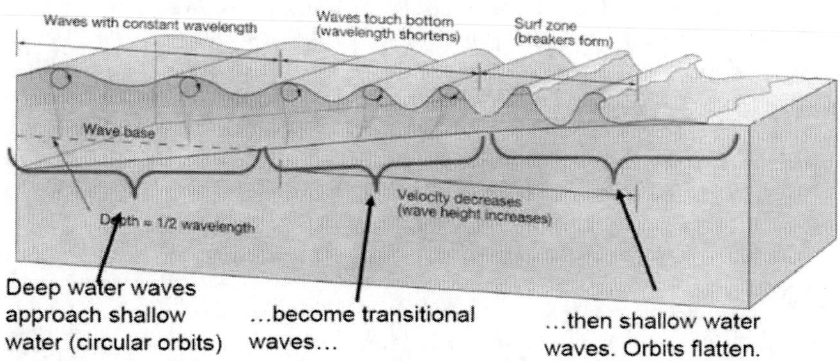

Figure 4. The process of tsunami propagation from deep water to shallow water (Zhang et al., 2020).

Conclusion and Implications

Current studies of the tsunami are incomplete. Numerical and experimental simulation is challenged due to the randomized tsunami nature. Therefore, the use of probabilistic methods to predict maximum tsunami highlands is essential and useful. The close agreement showed that the probabilistic model for the maximum tsunami height in this research was valid and rational. In this randomized model, the most important factor affecting the height of the tsunami was considered. The major potential consequences of the uncertainty during the release of a tsunami should be included in the analysis of the risk of a future tsunami. This helps to predict the risks of tsunamis to advance existing research.

References

Adams, L. M., LeVeque, R. J. and González, F. I., 2015. The pattern method for incorporating tidal uncertainty into probabilistic tsunami hazard assessment (PTHA). *Natural Hazards*, 76(1), pp.19-39.

Blaser, L., Krüger, F., Ohrnberger, M. and Scherbaum, F., 2010. Scaling relations of earthquake source parameter estimates with special focus on subduction environment. *Bulletin of the Seismological Society of America*, 100(6), pp.2914-2926.

Burbidge, D., Mueller, C. and Power, W., 2015. The effect of uncertainty in earthquake fault parameters on the maximum wave height from a tsunami propagation model. *Natural Hazards and Earth System Sciences*, 15(10), pp.2299-2312.

Davies, G., Griffin, J., Løvholt, F., Glimsdal, S., Harbitz, C., Thio, H.K., Lorito, S., Basili, R., Selva, J., Geist, E. and Baptista, M.A., 2018. *A Global Probabilistic Tsunami Hazard Assessment from Earthquake Sources.* Geological Society, London, Special Publications, 456(1), pp.219-244.

Davies, G., Horspool, N. and Miller, V., 2015. Tsunami inundation from heterogeneous earthquake slip distributions: Evaluation of synthetic source models. *Journal of Geophysical Research: Solid Earth*, 120(9), pp.6431-6451.

Dawson, A. G., Shi, S., Smith, D. E. and Long, D., 1993. Geological investigations of tsunami generation and long-term tsunami frequency. In *Natural Disasters: Protecting Vulnerable Communities: Proceedings of the Conference Held in London*, 13-15 October 1993 (pp. 140-146). Thomas Telford Publishing.

Dong, S., Zhai, J. and Tao, S., 2017. Long-term statistics of extreme tsunami height at Crescent City. *Journal of Ocean University of China*, 16(3), pp.437-446.

Geist, E. L., 2009. Phenomenology of tsunamis: Statistical properties from generation to runup. In *Advances in Geophysics* (Vol. 51, pp. 107-169). Elsevier.

Geist, E. L. and Lynett, P. J., 2014. Source processes for the probabilistic assessment of tsunami hazards. *Oceanography*, 27(2), pp.86-93.

Geist, E. L. and Parsons, T., 2006. Probabilistic analysis of tsunami hazards. *Natural Hazards*, 37(3), pp.277-314.

Glimsdal, S., Løvholt, F., Harbitz, C. B., Romano, F., Lorito, S., Orefice, S., Brizuela, B., Selva, J., Hoechner, A., Volpe, M. and Babeyko, A., 2019. A new approximate method for quantifying tsunami maximum inundation height probability. *Pure and Applied Geophysics*, 176(7), pp.3227-3246.

Glimsdal, S., Pedersen, G. K., Harbitz, C. B. and Løvholt, F., 2013. Dispersion of tsunamis: does it really matter? *Natural Hazards and Earth System Sciences*, 13(6), pp.1507-1526.

Goda, K., Mai, P. M., Yasuda, T. and Mori, N., 2014. Sensitivity of tsunami wave profiles and inundation simulations to earthquake slip and fault geometry for the 2011 Tohoku earthquake. *Earth, Planets and Space*, 66(1), pp.1-20.

González, F. I., LeVeque, R. J. and Adams, L. M., 2013. Probabilistic Tsunami Hazard Assessment (PTHA) for Crescent City, CA. *Final Report for Phase I*. University of Washington Department of Applied Mathmatics.

Greenslade, D. J., Allen, S. C. and Simanjuntak, M. A., 2011. An evaluation of tsunami forecasts from the T2 scenario database. *Pure and Applied Geophysics*, 168(6), pp.1137-1151.

Greenslade, D. J. and Titov, V. V., 2008. A comparison study of two numerical tsunami forecasting systems. *Pure and Applied Geophysics*, 165(11), pp.1991-2001.

Heintz, J. A. and Robertson, I.N., 2008. Design of structures for vertical evacuation from tsunamis. In *Solutions to Coastal Disasters* 2008: Tsunamis (pp. 72-81).

Kaiser, G., Scheele, L., Kortenhaus, A., Løvholt, F., Römer, H. and Leschka, S., 2011. The influence of land cover roughness on the results of high resolution tsunami inundation modeling. *Natural Hazards and Earth System Sciences*, 11(9), pp.2521-2540.

Kamigaichi, O., 2011. Tsunami forecasting and warning. *Extreme Environmental Events Complexity in Forecasting and Early Warning*.

Kienle, J., Kowalik, Z. and Murty, T. S., 1987. Tsunamis generated by eruptions from Mount St. Augustine volcano, Alaska. *Science*, 236(4807), pp.1442-1447.

Lee, E., Jung, T. and Shin, S., 2020. Numerical and probabilistic study on the optimal region for tsunami detection instrument deployment in the Eastern Sea of Korea. *Applied Sciences*, 10(17), p.6071.

LeVeque, R. J., George, D. L. and Berger, M. J., 2011. Tsunami modelling with adaptively refined finite volume methods. *Acta Numerica*, 20, pp.211-289.

Lin, S. C., Wu, T. R., Yen, E., Chen, H. Y., Hsu, J., Tsai, Y. L., Lee, C. J. and Philip, L. F. L., 2015. Development of a tsunami early warning system for the South China Sea. *Ocean Engineering*, 100, pp.1-18.

Løvholt, F., Pedersen, G., Bazin, S., Kühn, D., Bredesen, R.E. and Harbitz, C., 2012. Stochastic analysis of tsunami runup due to heterogeneous coseismic slip and dispersion. *Journal of Geophysical Research: Oceans*, 117(C3).

Mai, P. M. and Beroza, G. C., 2002. A spatial random field model to characterize complexity in earthquake slip. *Journal of Geophysical Research: Solid Earth*, 107 (B11), pp.ESE-10.

McCloskey, J., Antonioli, A., Piatanesi, A., Sieh, K., Steacy, S., Nalbant, S. S., Cocco, M., Giunchi, C., Huang, J. D. and Dunlop, P., 2007. Near-field propagation of tsunamis from megathrust earthquakes. *Geophysical Research Letters*, 34(14).

McCloskey, J., Antonioli, A., Piatanesi, A., Sieh, K., Steacy, S., Nalbant, S., Cocco, M., Giunchi, C., Huang, J. and Dunlop, P., 2008. Tsunami threat in the Indian Ocean from a future megathrust earthquake west of Sumatra. *Earth and Planetary Science Letters*, 265(1-2), pp.61-81.

Melgar, D., Allen, R. M., Riquelme, S., Geng, J., Bravo, F., Baez, J. C., Parra, H., Barrientos, S., Fang, P., Bock, Y. and Bevis, M., 2016. Local tsunami warnings: Perspectives from recent large events. *Geophysical Research Letters*, 43(3), pp.1109-1117.

Melgar, D. and Bock, Y., 2013. Near-field tsunami models with rapid earthquake source inversions from land-and ocean-based observations: The potential for forecast and warning. *Journal of Geophysical Research: Solid Earth*, 118(11), pp.5939-5955.

Melgar, D., Crowell, B. W., Geng, J., Allen, R. M., Bock, Y., Riquelme, S., Hill, E. M., Protti, M. and Ganas, A., 2015. Earthquake magnitude calculation without saturation from the scaling of peak ground displacement. *Geophysical Research Letters*, 42(13), pp.5197-5205.

Meyers, R. A. ed., 2009. *Encyclopedia of Complexity and Systems Science (Vol. 9)*. New York: Springer.

Mofjeld, H. O., González, F. I., Titov, V. V., Venturato, A. J. and Newman, J. C., 2007. Effects of tides on maximum tsunami wave heights: Probability distributions. *Journal of Atmospheric and Oceanic Technology*, 24(1), pp.117-123.

Mulia, I. E., Gusman, A. R. and Satake, K., 2018. Alternative to non-linear model for simulating tsunami inundation in real-time. *Geophysical Journal International*, 214(3), pp.2002-2013.

Murotani, S., Miyake, H. and Koketsu, K., 2008. Scaling of characterized slip models for plate-boundary earthquakes. *Earth, Planets and Space,* 60(9), pp.987-991.

Murotani, S., Satake, K. and Fujii, Y., 2013. Scaling relations of seismic moment, rupture area, average slip, and asperity size for M~ 9 subduction-zone earthquakes. *Geophysical Research Letters*, 40(19), pp.5070-5074.

Myers, E. P. and Baptista, A. M., 2001. Analysis of factors influencing simulations of the 1993 Hokkaido Nansei-Oki and 1964 Alaska tsunamis. *Natural Hazards*, 23(1), pp.1-28.

Popinet, S., 2015. A quadtree-adaptive multigrid solver for the Serre–Green–Naghdi equations. *Journal of Computational Physics*, 302, pp.336-358.

Ren, Z. Y., Zhao, X., Wang, B. L., Dias, F. and Liu, H., 2017. Characteristics of wave amplitude and currents in South China Sea induced by a virtual extreme tsunami. *Journal of Hydrodynamics*, 29(3), pp.377-392.

Rudloff, A., Lauterjung, J., Münch, U. and Tinti, S., 2009. Preface "The GITEWS Project (German-Indonesian Tsunami Early Warning System)." *Natural Hazards and Earth System Sciences*, 9(4), pp.1381-1382.

Setiyono, U., Gusman, A. R., Satake, K. and Fujii, Y., 2017. Pre-computed tsunami inundation database and forecast simulation in Pelabuhan Ratu, Indonesia. *Pure and Applied Geophysics*, 174(8), pp.3219-3235.

Shi, F., Kirby, J. T., Harris, J. C., Geiman, J. D. and Grilli, S.T., 2012. A high-order adaptive time-stepping TVD solver for Boussinesq modeling of breaking waves and coastal inundation. *Ocean Modelling*, 43, pp.36-51.

Strasser, F. O., Arango, M. C. and Bommer, J. J., 2010. Scaling of the source dimensions of interface and intraslab subduction-zone earthquakes with moment magnitude. *Seismological Research Letters*, 81(6), pp.941-950.

Thio, H. K., Somerville, P. and Ichinose, G., 2007. Probabilistic analysis of strongground motion and tsunami hazards in Southeast Asia. *Journal of Earthquake and Tsunami*, 1(02), pp.119-137.

Wells, D. L. and Coppersmith, K. J., 1994. New empirical relationships among magnitude, rupture length, rupture width, rupture area, and surface displacement. *Bulletin of the seismological Society of America*, 84(4), pp.974-1002.

Wu, T. R. and Huang, H. C., 2009. Modeling tsunami hazards from Manila trench to Taiwan. *Journal of Asian Earth Sciences*, 36(1), pp.21-28.

Yeh, H., Barbosa, A. R., Ko, H. and Cawley, J. G., 2014. Tsunami loadings on structures: Review and analysis. *Coastal Engineering Proceedings*, 1(34), p.4.

Zhang, Y. and Cao, Y., 2015. A fuzzy quantification approach of uncertainties in an extreme wave height modeling. *Acta Oceanologica Sinica*, 34(3), pp.90-98.

Zhang, Y. and Lam, J. S. L., 2014. Non-conventional modeling of extreme significant wave height through random sets. *Acta Oceanologica Sinica*, 33(7), pp.125-130.

Zhang, Y. and Lam, J. S. L., 2015. Reliability analysis of offshore structures within a time varying environment. *Stochastic Environmental Research and Risk Assessment*, 29(6), pp.1615-1636.

Zhang Z, Zeng Y., 2018. Probabilistic distribution of maximum tsunami height based on stochastic process of water depth. In: Qian XD, Pang SD, Ong GP, Phoon KK (eds) *Proceedings, Sixth International Symposium on Reliability Engineering and Risk Management, 31 May–1 June 2018*. Research Publishing, Singapore. ISBN: 978-981-11-2726-7.

Part III: Energy Demands and Water Security (EDWS)

Chapter 9

Reduction of Water Leakage by Smart Water Technologies

Shalini Pathak[1], Mir Bintul Huda[2,*], M. A. Lone[3] and Saeid Eslamian[4]

[1]National Institute of Technology Srinagar,
Hazratbal, Srinagar Jammu and Kashmi, India
[2]Water Resources Management Centre,
National Institute of Technology Srinagar,
Hazratbal, Srinagar Jammu and Kashmi, India
[3]Civil Engineering Department,
National Institute of Technology Srinagar,
Hazratbal, Srinagar Jammu and Kashmi, India
[4]Department of Water Engineering, College of Agriculture,
Isfahan University of Technology, Isfahan, Iran

Abstract

Water is an essential requirement on this earth. But, with urbanization and the population explosion, this resource is under continuous stress. It is evident from the depleting water reserves and the increasing demands of water for various purposes. The changing climate has further put stress on the management practices of dealing with this natural resource that sets apart the earth from all other planets. The chapter focuses on smart water management techniques to reduce leakages in the water resources systems for optimal usage and minimum wastage. The various components and techniques dealing with smart water management have been discussed.

---

* Corresponding Author's Email: dr.mirhuda@gmail.com.

In: Handbook of Uncertainty in Eurasian Forecasting (HEF)
Editors: Saeid Eslamian, Mir Bintul Huda and Faezeh Eslamian
ISBN: 978-1-68507-966-6978

Keywords: water, smart water management, leakage, sustainable development, climate change, efficiency

Introduction

Water is a basic human need that is required in sufficient quantity as well as specific quality and is essential for all aspects of life and sustainable development. Water resources is part of all forms of development including food security, health, and poverty elimination even in sustaining economic growth for agriculture, industry, and energy generation, and in the maintenance of a healthy ecosystem. Ecosystems related to water and the environment have always been natural sites for human settlements and civilizations, having various beneficial aspects such as transportation, irrigation, protection from floods, and habitats for biodiversity. The increasing population, intensification in the agricultural sector, urbanization, industrial production, pollution, and most importantly climate change are overwhelming and undermines nature's ability to provide its functions and services. Studies suggest that if the natural environment continues to be degraded and continuous unsustainable pressures are put on water resources 50 percent of the world's population with 40 percent of global agricultural production will be at risk by 2050. Populations most affected will be the poor and marginalized with further rising inequalities. Water management efficiency has become a major issue for many countries and water industries.

800 million people currently do not have access to clean drinking water. With events like floods, climate change, and water pollution, freshwater is getting scarcer. According to the World Bank database, developed countries have non-revenue water losses of 40% and because of these non-revenue losses, extra water must be pumped.

Unborn water operation requires immediate action including threat assessment with defense and relief systems that casts and has warning systems. Moreover, with the climate change impacts, extreme events will prove to be more frequent and severe than before. As a result, a necessity to adopt alternative strategies rather than the traditional approach focusing only on reducing water wastage and sustaining it for the future is required.

Smart Water Management involves the latest technologies, including, hardware, software, and analytics to help save water, reuse policies and targets of automation, and data gathering analysis in the water management sector. Smart water technology (SWT) has evolved for proper conduct and saving of water resources with technologies that can lead to better water resource operation, which will lead to the reduction of water deficiency worldwide. The advancement of Information Communication Technology (ICT) in recent times, particularly in the field of wireless communication, mobile communication affinity, cloud technologies, and computing hasled to the relinquishment of smart water operation technologies. All these advancements allow real-time monitoring of water architectures and their terrain, real-time analysis and predictions for early warning and decision-support systems, and fast response time to implicit flood or other water-related issues. The adaptability of smart water systems will give new tools to attack some of the problems that ultramodern metropolises are facing and foster innovative approaches to ensure safe and acceptable water provision. Smart Water Management (SWM) seeks to promote sustainable, well-coordinated development and operation of water resources through the integration of ICT products, tools and results therefore furnishing the base for a sustainable approach to water operation and consumption.

By employing the eventuality of ICTs in metropolises, the perpetration of SWM can help to overcome water-related socio-profitable, artistic, and environmental challenges, as well as equip metropolises with technology to alleviate the impacts of climate change. SWM is one of the many suggested options to tackle issues related to climate change by having a significant impact on the ecosystem and by reducing the water footprint.

Smart Water Management

The main goal of Smart Water Management is to sustain water security at every level possible in a sustainable and self-sufficient manner, through the help of information technology, monitoring and controlling various technologies, and the implementation of a system of all the processes required in a water cycle. A smart water system (SWS) makes use of sensors, data, and information communication technology (ICT) for providing real-time monitoring of pressure, the flow of water, quality of water, and moisture. Its capacity of detecting abnormalities ranges from losses due to non-revenue water to possible water contamination in the water distribution system.

Although rigorous planning with manual monitoring and various optimization tools have been developed and already implemented such as hydrologic models and real-time monitoring and control systems for treatment and distribution of water, reservoirs with decision support systems and hydraulic infrastructure, they usually do not communicate with each other and raises concerns to have a framework that can integrate and interlink all these applications in it together. ICT scans provide means for improving the productivity as well as efficiency of water management, to increase the submission of resources using advanced information technologies for observation, storage, process, and analyses of the system monitoring the data and presentation of the result analysis. The increased availability of a more intelligent, ICT-enabled approach for the management of water hassled the development of smart water management. Many countries have already incorporated information technology in a broad manner of policies ranging from Water Management to various fields such as electricity, transportation, and the environment.

The primary functions of water management are pressure management and smarter leakage techniques for networks of water, smart network operations and maintenance of the water cycle with a smart water meter, and smart water monitoring of quality. Countries around the world are getting increasingly more aware of the critical stage of the situation of water. The implementation of SWM will provide significant improvement in water distribution through detection of leakage and reduction, lower maintenance costs with the help of real-time monitoring and forecasting, and will be beneficial in improving customer service through real-time feedback on water usage and availability. It also benefits the water utilities economically by the detection of the connections and water theft.

Smart Water Management Framework

The main components include:

1. Sensors.
2. Monitoring and acquisition of data.
3. Conversion of data.
4. Processing and storage of data.
5. Support system and applications with monitoring and control system.

Different sensors with microcontrollers are embedded and through that, a system is created using the cloud services for the storage of data. The distribution of water across the country is uneven and requires an automated system that should be developed so that the water distributed is equal in pressure, even for checking different parameters of water quality like Ph, temperature, and turbidity it is of use. For checking the water level in tanks level sensors are used, and pressure and flow sensors are used to detect the leakage in pipes. The water level sensor device is designed in such a way that it can detect and even control the level of water in a water tank. The system at first senses the volume of water that is available in the tank through the flow of water and by the domestic pipeline it can be monitored with forecasting and visualization from any part of the world using the internet and through a laptop or a smartphone. The data collected are analyzed and predictions are made, and the users are informed about demand, asset, and leakage management. The system has already been tested and has proven to be successful. The work can be used to forecast data for the bigger community with customer validation for the better performance of the overall system. The model is used to forecast and monitor the water resources basically with the help of a flow meter, controller, and cloud infrastructure. This can also be possibly implemented to reuse the water and with the prevention water wastage. With continuous work, the proposed system can be made fully autonomous in the future by embedding it with AI and with some predefined rules. Artificial intelligence can help smart water systems to be carried out automatically without any human intervention.

Recommendations

Key steps for the integration of a smart water management system:

1. Adoption.
2. Implementation.
3. Operation and Maintenance.

The first step is the assurance of the specifications for interoperability and the data shared across the services which can be achieved through the development of tools that will properly contextualize the processes in the business and the water domain by making use of existing sensors and creating new water-specific ontologies for them that are not available.

The second step is the promotion of cross-domain sharing of data by developing and providing data interfaces that are open and establishing policies.

The third step is to get the support for successful completion of projects on smart water technology. The outcomes of these projects will be the contributing factor to the establishment of rules and guidelines for the best practices in the future for the implemented smart water technologies.

The implementation of Smart Water technology and its success requires the development of systems of standards for communities ensuring the interoperability of solutions. Both the inter systems that are within the water system and the system itself must not only be operated individually within its standards but must also interact with other specific systems.

Key performance indicators for SWM:

1. Development of a framework that can be benchmarked and adoption of assessment methodology.
2. Development of a supporting benchmarking tool.
3. Testing and validating the framework.
4. Availability of tools to the water sector.

Leakages within the water system distribution are caused by deterioration of the pipes through wear and tear, external pipe corrosion, or mechanical damage caused by external loads on the pipes, like earthquakes, excavation works, contraction and expansion of pipes thanks to temperatures extremes. Typical leakage detection methods often employ the utilization of acoustics surveys by monitoring the vibration generated by the water struggling to escape through a leak site. Recent developments have included installing pressure and noise sensors to watch and detect such changes in pressure and vibrations.

Challenges and Limitations of Smart Water Management

The major issue impeding the Smart Water Management are:

1. High cost for implementation and open & big data management.
2. Information System management of water resources.
3. Smart sensors technology.

Real-time data integration and monitoring give the users updated information for quick decisions, therefore, reducing losses. This is especially useful where work maintenance is being done and dispatch of the maintenance crew to the exact location of the incident can be done in minimal time. Data availability especially open data is a great resource that is not being tapped. Individuals and many organizations collect a different range of different types of data to perform their tasks. Data privacy and security are the major concerns. Security and safety of public data are important for intelligent sensors because they are the elements of wireless networks that can communicate directly or indirectly with each other. Thus, it is of prime importance that the anonymity of data and its assurance are secured when they are being shared.

For the water management and information system, there are various models and business utilities available with many of them coming from software facilities with the main objective being selling their software package solutions and not focusing on creating and ensuring interoperability. Therefore, the need for an overall approach that is consistent is being made for an integrated model of the water that involves multiple steps. One of the main challenges faced by the utilities and the concerned government for smart sensors technology is the decision of implementing smart sensors and deciding what type of technology is required for their water systems. Factors affecting the implementation: Cost of implementation, size of the infrastructure, software development, and real-time status of the data system.

Conclusion

The efficiency of water management through smart water management is being investigated and the available technologies and methodologies are being analyzed thoroughly. Smart Water Management combines existing techniques with Information Communication Technology to enable a more sustainable water management. The main component of the SWM is the implementation of the smart sensors where data are collected and are being transmitted in real-time to a controlled system that is centralized, and which requires a wide range of technologies ranging from large network communications to microelectronics to big data analytics. Analysis of the current smart water implementations has proven that open &big data management, information system management of water resources, and smart sensors technology as major concerns for the challenges and limitations. Leak detection indicators are being proposed and implemented by computing the effectiveness of the indicator by looking at the difference between before and after the occurrence of the leak and its validation using IoT. Future improvements can be achieved by redefining the current system and will be able to provide a more accurate analysis of the occurrence of the leak. Models that are statistical like artificial neural networks can be made to train and deal with real data of different scenarios of different leakages of the network to forecast the difference in flow. With sensor technology, the study of aquatic ecosystems have become faster and easier.

References

Abell, R., et al. (2017). *Beyond the Source: The Environmental, Economic and Community Benefits of Source Water Protection*. The Nature Conservancy, Arlington, VA, USA.

Agarwal, A., et al. (2000). Integrated Water Resources Management. *Global Water Partnership Technical Advisory Committee (TAC), Background Paper No. 4*.

Gupta, A., Mishra, S., Bokde, N. & Kulat, K.D. (2016). Need of smart water systems in India. *Int. J. Appl. Eng. Res.*, 11, 2216–2223.

Gupta, A. D., Pandey, P., Feijoo, A., Yaseen, Z. M. & Bokde, N. D. (2020). smart water technology for efficient water resource management: A review. *Energies,* 13(23), 6268. https://doi.org/10.3390/en13236268.

Liemberger, R., Wyatt, A. (2018). Quantifying the global non-revenue water problem. *Water Science and Technology,*19, 831–837. https://doi.org/10.2166/ws.2018.129.

Shiva Kumar, K.R., Gokul Manikanta, T., Vinod, P., Ramarao, N. & Girija, S. (2020). Design & development of water management system. *International journal of engineering research & technology(IJERT),* 09(03). http://dx.doi.org/ 10.17577/ IJERTV9IS030163.

Chapter 10

Potential of Solar Energy in India

Mir Bintul Islam[1], Mir Bintul Huda[2,*], Nasir Ahmad Rather[3] and Saeid Eslamian[4]

[1]Department of Electronics Engineering,
Jamia Millia Islamia, Delhi, India
[2]Water Resources Management Centre,
National Institute of Technology,
Srinagar, India
[3]Department of Civil Engineering,
BGSBU Rajouri, Jammu and Kashmir, India
[4]Department of Water Engineering,
College of Agriculture,
Isfahan University of Technology, Isfahan, Iran

Abstract

Solar Energy is the most readily available and practically inexhaustible power source that has helped in replacing conventional fossil fuels. It has started playing a major role in the sustainable development of the nation along with eliminating the adverse impacts on the environment. Because of its numerous advantages, most of the developed nations have switched over to it on a magnanimous scale. Following this trend, India has also done its due. Because of the geographical location of the country, solar energy has huge potential for development in this region, and many hotspots have been identified where this energy can be tapped efficiently. The chapter reviews the scope, potential, and steps taken to date by the government in this sector especially the Jawaharlal Nehru National Solar Mission, which has proved to be a guiding light on this path of sustainable development.

* Corresponding Author's Email: dr.mirhuda@gmail.com.

In: Handbook of Uncertainty in Eurasian Forecasting (HEF)
Editors: Saeid Eslamian, Mir Bintul Huda and Faezeh Eslamian
ISBN: 978-1-68507-966-6978

Keywords: solar, renewable energy, India, JNNSM, urbanization, energy demands, power deficit, potential

Introduction

With the ever-increasing urbanization and the population explosion throughout the world, the energy demand is on a steep rise. The world is dealing with a continuous crisis-like situation of competition for production and sustainable management of energy. Energy consumption has increased by 1.5% globally, the main contribution to this figure comes from the nations such as India, China, and Brazil where the population rise is substantially high (IEA, 2014). The main sources of energy are fossil fuels which account for about 79% of the total world energy consumption (IEA, 2006). With the increasing population, these fossil fuels are under great stress and their depletion is inevitable at a very quick pace, this has led researchers and nations on shifting the focus to alternative fuels which can ensure sustainable development. This scenario is the same throughout the globe including India.

India is a developing nation and in terms of its population is placed in the second position in the world. The population, industrialization, urbanization, and modernization are on a steep rise which is the reason for the enormous power requirement in this region. Most of the power generation plants are dependent on fossil fuels and at present, the capacity of these plants stands at 27122 MW which accounts for about 69.5% of the total power generated in the country (MEP, 2016). This is a big figure and will have greater implications on fuel sustainability, economic feasibility, and environmental impacts. Fossil fuels are often coupled with greenhouse gas emissions on a large scale. Renewable energy sources can solve the problem of sustainable development associated with fossil fuel-based power plants as these energy sources are unlimited, eco-friendly, and provides energy with negligible emissions of air pollutant and greenhouse gases (Varuna and Singhal, 2007). There are various forms of alternate energy sources viz. solar, wind, geothermal, ocean, etc. But most of these sources have their limitations, they are restricted to certain sites and places, and their potential is restricted except for solar energy, which is available in all places and is unlimited. The location of India on the globe makes it an ideal region, where solar energy can be utilized efficiently and can replace fossil fuels substantially. This will help in reduce greenhouse gas emissions and help in achieving sustainable growth. India receives about 5000 trillion kilowatt-hours (KWh) of theoretically calculated solar energy per year (Muneer et al. 2005), but still, the solar power contribution is just about 2.5% of the total installed capacity of power stations which accounts for only 7.8 GW of totaled installed capacity. In 2015, the peak power deficit in India was about 2.6%, this is expected to rise to 5.6% in 2022 because of the population rise and economic growth (ASSOCHAM, 2016). To maintain sustainable growth of 8% in the gross domestic product (GDP) the electricity sector must grow at a rate of 7%. To provide electricity to 1560 million people by 2035, an additional installed capacity of 450 GW is required. Solar energy is an ideal way to fulfill the power deficit in our country, the geographical location of India constitutes deserts, hills, plateaus, forests, etc.; so it can provide an economical solution to power demand and consumption (CEA, 2015).

Status of Solar Energy in India

India lies between 8°4'to 37°6'N and 68°7'to 97°25'E. It is the seventh-largest country in terms of land having about 2.9 million square kilometers, making its potential for solar energy development huge. In 1980, India was the first nation to establish a Ministry of non-conventional energy resources. Out of all the non-conventional resources, wind energy and solar energy can be termed truly renewable, environment friendly, and freely available. Further solar energy is still superior to all since the wind energy tapping is limited to the wind speed, the location where it can be tapped, and unpredictable. But the tapping of solar energy is the easiest and can be exhaustively incorporated at all levels. Its continuous occurrence at all places is a boon and can help in overcoming the power deficit of the nation.

The solar radiation incident over India is equal to 4-8 kWh per square meter per day with annual radiation ranging from 1200-2300 kWh per square meter. It has an average of 250–300 clear sunny days and 2300-3200 hours of sunshine per year. India's electricity needs can be met on a total land area of 3000 km^2 which is equal to 0.1% of the total land in the country. The annual average direct normal irradiance in most of the Indian states is around 4.5 to 5.0 $KWh/m^2/Day$ and the average global horizontal irradiance is around 5.0 to 5.5 $KWh/m^2/Day$ (NREL, National Renewable Energy Laboratory), it is sufficient to produce 6,081,709 TWh/year (USDE). There are regions that receive solar insolation of more than 5 $kWh/m^2/day$ and can constitute solar hotspots in India such as the Western Ghats, the Eastern ghats, Gangetic plains plateau, Thar deserts, and Gujarat plains (Ramachandra et al. 2011).

The National Solar Mission is a major initiative of the Government of India and State Governments to promote ecologically sustainable growth while addressing India's energy security challenge. It will also constitute a major contribution by India to the global effort to meet the challenges of climate change. The National Action Plan on Climate Change also points out: "India is a tropical country, where sunshine is available for longer hours per day and in great intensity. Solar energy, therefore, has great potential as a future energy source. It also has the advantage of permitting the decentralized distribution of energy, thereby empowering people at the grassroots level." Based on this vision National Solar Mission was launched under the brand name "Solar India" (Veeraboina and Ratnam 2014).

The Jawaharlal Nehru National Solar Mission (JNNSM), or the National Solar Mission, is an initiative of the Government of India and State Governments to promote solar power in India. It was inaugurated in 2010 and has been revised twice and now targets 100 GW of solar PV (photovoltaic) by 2022. The objective of JNNSM is to establish India as a global leader in solar energy by creating the policy conditions for its deployment across the country. The Mission adopted a 3-phase approach, (2012-13) as Phase 1, the remaining 4 years of the 12th Plan (2013-17) as Phase 2, and the 13th Plan (2017-22) as Phase 3.

Off-Grid Solar PV Programme

National Solar Mission targeted 2000 MWp for off-grid solar PV applications. Under Phase-I (2010-13), a target of 200 MWp was kept against which 253 MWp was sanctioned, and under Phase-II (2013-17), a target of 500 MWp was kept against which 713 MWp has been sanctioned. Under Phase-III, a target of 118 MW has been kept excluding solar pumps which are to be installed under the PM KUSUM Scheme, and solar home lights which are being taken up under the 'Saubhagya' Scheme of the Ministry of Power. The detailed status of the off-grid and Decentralised Solar PV Applications Programme is given in Table 1.

Solar pumps are an important component of this program as it provides reliable irrigation facility in rural areas. It can meet the irrigation requirements for small and marginal farmers and can easily replace the existing diesel pumps on a large scale. It was first started by MNRE in 1992. 11600 solar pumps were installed in 1992-2014, in 2014-2015, Rs 400 crores were set aside for the installation of one lakh solar pumps. To date 2.37 lakh solar pumps have been installed.

Table 1. Status of the off-grid and Decentralised Solar PV Applications

System	No. of units/capacity installed
Solar Lamps/Lanterns	65,17,180
Solar Pumps	2,37,120
Solar Street Lights	6,71,832
Solar Home Lighting Systems	17,15,639
Solar Power Plants/Packs	212 MWp

Source: MNRE, GOI.

The government has recently launched a new scheme named Pradhan Mantri Kisan Urja Suraksha evam Utthan Mahabhiyan (PM KUSUM) which aims to install new standalone solar pumps in off-grid areas and solarize existing grid-connected agricultural pumps. This will provide farmers with a reliable source of irrigation and increase farmers' income and their overall economic status.

Ministry is also implementing a scheme for 70 lakh Solar Study lamps, which aims to provide rural students with high-quality and affordable clean light. Under Atal Jyoti Yojana (AJAY), Solar LED Street Lights in rural, semi-urban and urban areas are being installed in various States.

Grid-Connected Programme

The solar power sector in India has advanced at a fast pace. It helps in achieving sustainable growth as well as emerging as an integral part of the solution to meet the nation's energy needs and is an essential player for energy security. By 2022, a target of 100 GW of grid-connected power has been set up. To achieve it, GOI has launched various schemes like Solar Park Scheme, VGF Schemes, CPSU Scheme, Defence Scheme, Canal bank and Canal top Scheme, Bundling Scheme, Grid Connected Solar Rooftop Scheme, etc. Schemes and Policies are being taken up to promote grid-connected solar power plants.

Recently, India has achieved 5th rank in the world in solar power deployment. Solar power capacity has increased by more than 11 times in the last five years from 2.6 GW in March 2014 to 28.18 GW in March 2019. With technological improvements, the economy of scale, and a reduction in solar cell/ module prices, solar tariff in India is now competitive and has achieved grid parity.

Table 2. Grid-connected solar power projects in India

S. No.	Year	Capacity added (MW)	Cumulative capacity (MW)
1	Up to 2010	8.54	11.35
2	2010-2011	24.58	35.93
3	2011-2012	896.37	932.30
4	2012-2013	752.16	1684.46
5	2013-2014	947.46	2631.93
6	2014-2015	1112.07	3743.97
7	2015-2016	3018.88	6762.87
8	2016-2017	5525.98	12288.85
9	2017-2018	9362.63	21651.48
10	2018-2019	6529.20	28180.68

Solar power is the most promising solution for dealing with the energy crisis globally, especially in developing nations such as India, which requires a proper balance between economic progress, environment, sustainable development, and the energy crisis. Solar PV technology will change the world's energy dynamics. It will be very crucial in the economic development of India. With a meager 10 MW solar energy generation in 2010, the figure has risen to 30GW in 2019. The steps taken by the government in this sector have helped in its boost. The growth of this sector is dependent on several parameters; population, public awareness, investment, returns, environmental considerations, development in the rural sector, cost-effectiveness, etc.

Government Initiatives

The government of India has invested in this sector on a large scale. Since 2000, the various regulations/policies adopted by the government are as follows:

- Electricity Act 2003
- National Electricity Policy 2005
- National Tariff Policy 2006
- National Rural Electrification Policies (NREP), 2006
- Tariff policy (January 2006)

- Initiatives to promote solar PV in India
- Semiconductor Policy (2007)
- National action plan on climate change (NAPCC) – 2008
- National five-year plan (FYP)
- Jawaharlal Nehru National Solar mission

National Action Plan on Climate Change (NAPCC)

GOI developed NAPCC in 2008 for achieving sustainable growth and development in the energy sector. The plan was developed to address the issues of environment and climate change on the national level. It aimed at reducing the contribution to global greenhouse gas emissions. One of the eight missions of this plan was related to solar power development in India via JNNSM. According to NAPCC, 15% of India's energy could come from renewable sources by 2020 and out of this, the major portion is from solar power.

Solar Pumping Program

This program was launched in 1992. About 13964 solar pumps were installed from 1992-to 2014. In those times, the cost of installation of solar panels was high, but with the efforts of the government, the sector got uplifted. With time the solar panel installation and the related capital cost got rapidly reduced which led to a substantial increase in the number of solar pumps installed for irrigation and drinking purposes. States are identifying the potential solar pump targets for determining the target number of pumps to be replaced by the solar pumps, especially in the agricultural sector.

Solar Parks

Solar Parks are the regions solely dedicated to solar power development. It requires a one-time capital investment which is quite high. These solar parks will have all types of basic amenities required for setting up a solar power project like transmission lines, land, and water. For setting up solar power projects inside these Solar Parks, the developer doesn't require any type of clearance or approval from local/federal authorities. This step will encourage national and international developers/investors to set up their solar power projects in India.

Ministry of New and Renewable Energy (MNRE) has initiated a scheme for setting up 25 Solar Parks, each with a capacity of 500 MW and above.

Table 3. Setting up of solar parks

S No.	State	Capacity of solar park (approx)	Land famous at
1	Gujarat	750 MW Taluka- Vav, Distt.- Banaskantha	750 MW Taluka- Vav, Distt.- Banaskantha
2	Madhya Pradesh	750 MW +750 MW	Rewa-Distt
3	Telangana	1000 MW	Mehboob Nagar-Distt.
4	Andhra Pradesh	2500 MW	Anantpur-Distt, Kadapa and Kurnool
5	Karnataka	1000 MW	Mulwar, Bijapur (near Kargi)
6	Uttar Pradesh	600 MW	District-Jalaun, Sonbhadra, and Allahabad
7	Meghalaya	50 MW	University of Science and Technology, 9thMile, Near Guwahati, Meghalaya
8	Jammu& Kashmir	7500 MW	Leh and Kargil
9	Punjab	1000 MW +1000 MW	District-wise land identified
10	Rajasthan i. Bhadla Phase-II ii. Bhadla Phase-III iii. Jaisalmer ParkPhase-I iv. Jaisalmer ParkPhase-II	 700 MW 1000 MW 1000 MW 1000 MW	District - Bhadla and Jaisalmer
11	Tamil Nadu	500 MW	Location yet to be identified
12	Odhisa	1000 MW	Location yet to be identified
	Total	**22,100 MW**	

Solar Cities

The government has taken initiatives for the development of solar cities across the country in the year 2013-2014. The Solar City aims at a minimum 10% reduction in the projected demand for conventional energy at the end of five years, through a combination of enhancing supply from renewable energy sources in the city and energy efficiency measures.

Table 4. Approved solar cities (MNRE)

State/UT	Solar city	Pilot city	Model city
Andhra Pradesh	Kakinada Narsapur Town	Vijayawada	-
Assam	Guwahati Jorhat	- -	-
Arunachal Pradesh	Itanagar	-	-
Bihar	Gaya	-	-
Chandigarh	-	Chandigarh	Chandigarh
Chattisgarh	Bilaspur	Raipur	-
Gujarat	Surat	Rajkot	Gandhinagar
Goa	Panaji City	-	-
Haryana	Gurugram	Faridabad	-
Himachal Pradesh	Hamirpur	Shimla	-
Jammu & Kashmir	-	Leh	-
Karnataka	Hubli-Dharwad		Mysore
Kerala	Thiruvananthapuram Kochi	-	-
Maharashtra	Kalyan-Dombivli Aurangabad Nanded Pune	Thane Shirdi	Nagpur
Madhya Pradesh	Indore Gwalior Bhopal Jabalpur Rewa	-	-
Manipur	Imphal	-	-
Mizoram	-	Aizawl	-
Nagaland	Kohima Dimapur	-	-
Delhi	New Delhi (NDMC area)	-	-
Odisha	-	-	Bhubaneshwar
Punjab	Ludhiana SAS Nagar (Mohali)	Amritsar	-
Rajasthan	Ajmer Jaipur Jodhpur	-	-
Tamil Nadu	-	Coimbatore	-
Telangana	Mahbubnagar	-	-
Tripura	-	Agartala	-
Uttarakhand	Dehradun Haridwar and Rishikesh Chamoli-Gopeshwar	-	-

Table 4. (Continued)

State/UT	Solar City	Pilot City	Model City
Uttar Pradesh	Agra Moradabad Allahabad	-	-
West Bengal	Howrah Madhyamgram New Town Kolkata	-	-
Puducherry	-	Puducherry	-

The objectives of the program are:

- To empower the local government to address the energy challenges at the city level.
- To prepare a master plan for future demand by analyzing the current energy situation.
- To create awareness among people about solar energy through education and training.
- To involve stakeholders in the planning process.

A total of 60 cities/towns are proposed to be supported for development as Solar Cities. A city will be identified as a solar city based on the availability of renewable energy and population. Up to Rs. 50.00 Lakhs per city/town is provided depending upon population and initiatives decided to be taken by the City Council/ Administration. Among the 60 selected cities, Chandigarh, Bhubaneshwar, Gandhinagar, Nagpur, and Mysore have been identified as model solar cities as they have been doing notable work in promoting solar power.

Renewable Energy Certificates (REC) and Renewable Purchase Obligation (RPO)

The National Tariff Policy in January 2011 has prescribed solar-specific RPO to be increased from a minimum of 0.25 percent in 2012 to 8 percent by 2022. Central Electricity Regulatory Commission (CERC) and State Energy Regulatory Commission (SERCs) have issued various regulations, including solar RPOs, REC framework, tariff, grid connectivity, forecasting, etc. for promoting solar energy. The Renewable Energy Certificate (REC) mechanism is a market-based instrument to promote renewable energy and facilitate the compliance of renewable purchase obligations (RPO). It is aimed at addressing the mismatch between the availability of RE resources in the state and the requirement of the obligated entities to meet the renewable purchase obligation (RPO). Renewable Energy Certificates (RECs) are aimed at providing an easier avenue for various entities, including power distribution companies, to meet their green energy obligations. To make the Renewable Purchase Obligation (RPO) effective and in turn develop the market for renewable energy (particularly solar energy), the Central Energy Regulatory Commission (CERC) introduced RECs. A renewable energy generator under the REC scheme gets revenue from the sale of electricity and the revenue from the sale of environmental attributes in the form of RECs (Rathore et al. 2017).

Research Institutes

The National Institute of Solar Energy (Solar Energy Centre) is the apex National R&D institution in the field of Solar Energy. It was established by MNRE in 1982. The Government of India converted the 25-year-old Solar Energy Centre (SEC) under MNRE to an autonomous institution in September 2013 to assist the Ministry in implementing the National Solar Mission and to coordinate research, technology, skill development, training, consultancy, incubation, and other related works. In 2010, National Centre for Photovoltaic Research and Education (NCPRE) at IIT Bombay was launched as a part of JNNSM. NCPRE aims to create and execute the blueprint for human resource development for PV in India. The research includes silicon solar cell fabrication, characterization, energy storage, new materials, novel PV structures, power electronic interfaces for solar PV systems, new product designs, technology assessment, and environmental and social impact. Solar Energy Corporation of India Limited (SECI), was launched in 2011, it is a public sector undertaking, which is under the direct control of the Ministry of Renewable and Energy Resources to achieve the target of the National Solar Mission. It is responsible for the implementation of several schemes of MNRE, the major ones being the Viability Gap Funding (VGF) schemes for large-scale grid-connected projects under JNNSM, solar park scheme, and grid-connected solar rooftop scheme, along with a host of other specialized schemes such as defense scheme, canal-top scheme, Indo-Pak border scheme, etc.

Challenges

Because of the numerous advantages of this sector, there have been continuous development and innovations in it. Coupled with the enormous support from the Government, solar power development is reaching new heights and is playing a vital role in the sustainable development of the country. But many challenges are still being faced that need to be overcome to achieve all the set targets.

The initial installation cost and infrastructure are huge, which has greatly hindered in the adoption of this system at the societal level. Although various gadgets are being adopted by the urban households at the individual level, the societal level is very important, especially in the rural setup. Once this is achieved, it can help in the overall upliftment of the standard of living at the village level making them self-sufficient. The government at its level is trying to achieve the objectives, but the social contribution is a must, public awareness regarding the advantages and the willingness to shift to the new system is a must. Local NGOs and research institutes have a greater role to play. This will not just benefit the development of the nation but also help in sustainable development keeping the environmental considerations at par with globalization.

Conclusion

India is an emerging nation and is dealing with a steep population rise, development stress, and the crisis of meeting energy demands. All these coupled with the environmental impacts makes it mandatory to incorporate a sustainable development approach. With the fast-depleting fossil fuels and their disadvantages of greenhouse gas emission, non-conventional energy sources such as solar power have the potential for substantial development. The initial installation and replacement costs may be high, but the cost-benefit analysis lists the higher returns in a longer time frame. The chapter highlights the scope and the potential of solar energy in India, especially the hotspots that can be developed because of their advantages. It is interesting to note that the Government of India has incorporated this sector exhaustively in its policymaking and initiatives. A lot has been proposed and a substantial portion has been achieved by both off-grid programs as well as grid-connected sources. Solar Parks and Solar cities have been developed to achieve the targets set by the JNNSM. A continuous research and development approach has been adopted.

References

The Association Chambers of Commerce & Industry of India. http://www.assocham.org/newsdetail.php?id=5713.

Central Electricity Authority. http://cea.nic.in/reports/monthly/executivesummary/2015/exe_summary-08.pdf.

International Energy Agency (IEA). *Key world energy statics.* http://www.iea.org/Textbase/nppdf/free/2006/key2006.pdf.

International Energy Agency (IEA). *World Energy Statistics and Balances. 2014 Edition*, (Paris: OCED/IEA, 2014). https://www.iea.org/publications/freepublications/ publication/KeyWorld_Statistics_2015.

Ministry of New and Renewable Energy Source (MNRE). http://www.mnre.gov.in.

Ministry of Power, Government of India. *Power Sector at a Glance all India.* http://powermin.nic.in/content/power-sector-glance-all-india.

Muneer T, Asif M, Munawwar S. (2005). Sustainable production of solar electricity with particular reference to the Indian economy. *Renewable and Sustainable Energy Reviews*, 9 (5): 444. https://doi.org/10.1016/j.rser.2004.03.004.

Ramachandra, T. V., Rishabh Jain, and Gautham Krishnadas. Hotspots of solar potential in India. *Renewable and sustainable energy reviews* 15.6 (2011): 3178-3186.

Rathore, P. K. S., Srivastava, S. and Gupta, N. K. 2017. Solar energy in India: Key for sustainable development. *Proceedings of the International Conference on Frontiers in Engineering, Applied Sciences and Technology, NIT Tiruchurapalli, 2017.* Pages 22-27, ISBN 978-81-908388-8-7.

U.S Department of Energy. *Open Data Catlog.* http://en.openei.org/doe-opendata/dataset/solar-resources-by-class-and-country/resource/3e72f32a-7de1-4e5d-a25a-76928769625f.

Veeraboina P and Ratnam GY, Analysis of the opportunities and challenges of the solar water heating system (SWHS) in India: Estimates from the energy audit surveys & review. *Renewable and Sustainable Energy Reviews*, 16 (2012) 668– 676.

Chapter 11

Sustainable Management for Conservation of Urban Aquatic Forms in Bangladesh

Prabal Barua[1,*], Saeid Eslamian[2] and Anisa Mitra[3]

[1]Department of Knowledge Management for Development, Young Power in Social Action (YPSA), Chittagong, Bangladesh
[2] Department of Water Engineering, College of Agriculture, Isfahan University of Technology, Isfahan, Iran
[3]Department of Zoology, Sundarban Hazi Desarat College, West Bengal, India

Abstract

Bangladesh is facing a water crisis that has put stress on economic development, livelihoods, human well-being, and ecological sustainability at risk. In recent times, this country has witnessed rapid, unplanned urban expansion accompanied by serious environmental degradation. Conventional urban aquatic forms are known as ponds, canals, and any types of water bodies available in an urban city. The authors attempt to illustrate and examine the management of urban water bodies in the commercial capital city of Bangladesh by focusing on their characteristics and spatial distribution, using patterns and perceived health impacts on users and causes of disappearance and consequence of losses and degradation in this book chapter. The number of water bodies in the commercial capital city was found declining gradually. The authors found that the proper coordination of urban autonomous bodies, citizen awareness, community movement, the inclusion of urban management-

---

* Corresponding Author's Email: prabalims@gmail.com.

In: Handbook of Uncertainty in Eurasian Forecasting (HEF)
Editors: Saeid Eslamian, Mir Bintul Huda and Faezeh Eslamian
ISBN: 978-1-68507-966-6978

related laws, rules, and regulations, and legislative strengthening would reduce the negative impacts of unplanned urbanization in urban areas of Bangladesh. The authors have recommended effective implementation of the existing laws and regulations at the field for a sustainable and healthy environment for the inhabitants.

Introduction

Urban Water Bodies (UWBs) are a micro-morphological unit of the urban physical environment, most often referring to small accumulations of water, such as a pond, *dighi, doba, khal*, river, canal, lake, reservoirs, and humid fallow land in urban areas. Since freshwater constitutes just 0.01 percent of the total hydrosphere, and water stored in lakes, ponds and wetlands is a small fraction of it, the role of UWBs as the most productive life support systems can hardly be emphasized. As a micro-morphological unit, it has led to big consideration in urban areas. Especially, urban floodwater retention, erosion and sedimentation control demystification, microclimate, provision of habitat for organisms, and support of aquatic life. In contemporary towns and cities, waterfronts have remained the focus of many economic and recreational activities (Hassan et al., 2015).

The urban water supply system has turned out to be a more serious concern in the socio-economic situation of a nation and a very important feature in recent times, and specifies the quality of urban life (Banu, 2018). As an alternative, diverting rainwater in many small water harvesting structures in a catchment captures and stores more of the scarce rainfall closer to communities in arid parts of the world than a large reservoir downstream. General characteristics of small water bodies are types and numbers, area, water retention ability, year of creation, shape, depth ownership, and operation. Some common characteristics of small water bodies in Bangladesh are changes in the number of small water bodies, year of creation, process, and reason of creation, monitoring the size of SWB, shape, depth of water, ownership, operation, sludge, and water weeds and uses. Instead, specific characteristics of small water bodies are types, number, and percentage of SWBs, area, areal variation, shape, depth, location, association, water retention, seasonal variation, year of creation, causes of creation, watercolor, odor, submerged type, pollution rate, sludge and types of sludge, ownership, etc. (Barua et al., 2020; Barua and Rahman, 2020).

However, these have been subjected to gradual disappearance recently, including significant degradation of their water quality. Landscape alteration has a multiplying effect on biodiversity, water quality, and carbon emissions, with many other negative impacts on aquatic and terrestrial ecosystems. In recent decades, rapid urbanization has been accelerating land cover change in such a way that has an enormous impact on the radioactive, thermodynamic, and hydrological processes that can lead to a change in the local climate, in terms of, for example, temperature, precipitation, and cloudiness (Hasan et al., 2015; Alam et al., 2013).

Bangladesh is a developing country and one of the most populous countries in the world. In recent times, this country has witnessed rapid, unplanned urban expansion accompanied by serious environmental degradation. Owing to the accelerated urban growth, however, the area has experienced serious environmental degradation and several ecological problems. Nearly 56 percent of the land cover had changed, mainly because of the expansion of built-up areas and other human activities over 36 years along with the city expansion by 618%, with an average annual rate of increase of 17.5% (Kaufmann et al., 2017).

The rapid growth of population, urbanization, and industrialization creates tremendous pressure on land-use changes in the urban areas of Bangladesh. At present, Urban Water Bodies are in severe threat, disappearing, changing in size, depth, and shape rapidly due to natural and man-made activities. Attitudes towards UWBs are changing; these are no longer viewed as degraded systems of little value, but as more productive ecosystems with intrinsic values. The role of Urban Water Bodies in providing important regulating and cultural ecosystem services is also increasingly recognized. (Rahman, 2012).

In this book chapter, the author focuses on the sustainable management of urban aquatic forms in Chittagong city of Bangladesh. The Chittagong Metropolitan City is the second-largest metropolitan of Bangladesh and is recognized as the commercial capital city of the nation with geography that includes hills, plain lands, ponds, ditches, lakes, and other water bodies (Osmany, 2006). Due to this situation, citizens have been suffering from irregular, inadequate, and unsafe water supply due mainly because of inefficient management practices and this situation is worse in low-income residential areas. The limited supplied water was unsafe for household purposes, especially drinking because nearly 56% of the citizens have reported that their family members have exposure to different types of water-borne diseases, especially diarrhea, dysentery, cholera, typhoid, skin diseases at least once. Females and children are found to be more affected by the said diseases than their male counterparts. As a result, citizens, especially lower-class income- generating groups (urban poor: low income and slum people) directly used water from different water bodies, namely, digging ponds for household consumption purposes, and middle-class income-generating groups were practically used water from these water bodies. In contrast, water bodies in Chittagong city are disappearing fast due to a lack of awareness, renovation, and proper monitoring by the authorities concerned alongside the increasing demand for land for housing.

So, the objectives of this study are the exploration of the characteristics of the urban water bodies, pollutant status, and trend analysis of changing patterns of water bodies in the urban city. Lastly, the author developed sustainable management strategies for the conservation of the urban water environment for Chittagong city, which is the commercial capital city of Bangladesh.

Profile of the Chittagong City of Bangladesh

The study area Chittagong city which is known as the commercial capital city of the country is located between $22^0 15^/$ and $22^0 25^/$ North Latitudes and between $91^0 45^/$ and $91^0 55^/$ East Longitudes along the southeast coast of Bangladesh, bounded by the Karnaphuli River and its tributary the Halda River to the south and east, by the Bay of Bengal to the west and by a range of hills to the north. Chittagong city has various manufacturing establishments that extend about 10 miles (16 km) to the north and northeast of the urban city. This commercial capital city also has an export-processing zone and branch offices of many foreign firms and banks. The city is inhabited by over 5 million people and about 0.5 million people are served by 21,000 house connections, 0.2 million people collect water from the 588 street hydrants that are installed and maintained by the Chittagong Water Supply and Sewerage Authority (CWASA). The area's major rivers are the Karnaphuli, Halda, Sangu, and Matamuhari. The higher parts of the Chittagong Hills are forested, while the lower portions are covered predominantly with brushwood.

According to the world urbanization Perspective Report, Chittagong is the 56th urban agglomeration and the 10th fastest-growing city in the world. The busy seaport of the country makes the city the main international business hub, as well as the commercial capital of the country. The Chittagong Metropolitan area has a population of 5.00 million in 770 square kilometers of land area.

The Chittagong Metropolitan area has a population of 5 million in 770 square kilometers of land area. The rapid growth of population, urbanization, and industrialization created tremendous pressure on land-use changes in the city (Reza et al., 2017).

Importance of Urban Water Bodies

Chittagong City is the second metropolitan and commercial capital of Bangladesh, where different categories of stakeholders entrance for different purposes. Consequently, the city population is increasing day by day. Urban supply water is the basic amenities for city dwellers. City authority (CWASA) bounded to meet this demand for water for city dwellers. Unfortunately, CWASA didn't provide the demand for water for city dwellers. In this situation, city dwellers directly and partially depended on Urban Water Bodies (UWBs) to meet their daily needs, although this situation depends on geographical location, topographic conditions, the density of population, and consumption point's distance from the production center, income, or economic status and so on.

In this Chittagong city, most of the inhabitants opined that UWBs had been used as alternative water sources for meeting the daily water needs. City dwellers use this water for different household purposes such as water for the household, irrigation, vehicle washing, and restaurant using purposes. Besides, some citizens assumed that drinking water purpose is one of the major purposes of utilization of urban water bodies in Chittagong Metropolitan City. Subsistence agricultural activities particularly paddy and seasonal vegetables have been cultivated on the Northern side especially wards no 01 *(South Pahartali)*, 02 *(Jalalabad)*, 04 *(Chandgaon)*, and 05 *(Mohra)*, Eastern side particularly 18 *(East Bakalia)* and 35 *(Boxir Hat)*; Western and southern sides mainly 10 *(North Kattali)*, 11 *(South Kattali)*, 26 *(North Halishahar)*, 37 *(North Middle Halishahar)*, 38 *(South Middle Halishahar)*, 39 *(South Halishahar)*, 40 *(North Pothenga)* and 41 (*South Pothenga*) respectively. For agricultural purposes, water was used for irrigation purposes during the winter and summer seasons from UWBs.

Year of Origin of Urban Water Bodies and Utilisation

Urban territorial expansion is directly correlated with the creation of water bodies in the Chittagong Metropolitan city. Population agglomeration, the place for commercial, industrial, and residential purposes, the ground level changes of Chittagong city starting from 1948 through 1975 and finally in 1992. As a result, City Corporation, community, and personal landowners were digging a lot of water bodies for urban household purposes sometime industrial entrepreneurs dug ponds for emitting industrial waste and wastewater in CMC. Maximum water bodies were created almost in the nineteen centuries in the Chittagong Metropolitan city. It was remarkable that four major and well-known water bodies, namely *Agrabad Deva, Valuar dighi,* and *Jora dighi* were created during the establishment of the Bangladesh railway in 1853. In the checklist survey, citizens opined that few ponds existed since pre-history because they were hearing about existing water bodies appearing from their father, grandfather, and post-grandfather. During the year 2020-2021, a port connected road was built along with the beach from Chittagong port to Salimpur in Sitakunda. During the building of this road, several small water bodies were dug for the collection of soil for building the roads. It has increased the total number of water bodies in Chittagong Metropolitan City.

For functional activities, water bodies were used for multidimensional purposes in urban areas of Bangladesh. Middle and lower- class income-generating groups directly used water from urban water bodies for household purposes, sometimes drinking water collection from these water sources. Urban water bodies are the alternate water sources for every urban dweller because of the shortage and unsafe supply of Metropolitan piped water. There are 16 categories of urban water bodies were used in the Chittagong Metropolitan City. Maximum water bodies were used for household purposes, particularly bathing, clothing, and all types of households washing purposes. Although a large portion of the detected water bodies was used for household purposes together with fishing activities were done, these water bodies, separately second large portions of the water bodies were used for fishing purposes in the Chittagong Metropolitan City. Somewhat were used for dumping sewerage and conservation of water for causality purposes, that means fire brigade water.

In the book chapter, the authors found the 6 types of socio-cultural importance of urban water bodies in Chittagong city, particularly growing natural beauty in the house lane areas, road dividers, and surrounding environment. Occasionally, City Corporation, local schools, colleges, and other institutions were arranging swimming competitions in the city area that time UWBs were used for swimming competition purposes. Besides, in the late afternoon and evening after a day, respondents are back from working places, and they are sitting on the wall of the pond for resting, gossiping, and recreation purposes.

The environmental importance of Urban Water Bodies is very much essential in Chittagong City. Chittagong city is a commercial hub, industrial and port city, for these reasons rapid and unplanned urbanization, and unexpected industrial and residential development, reduce impervious space in the whole city area. In this point of view, UWBs have a significant role in friendly or safe dwelling city life. In the book chapter, the author mentioned that Urban Water Bodies (UWBs) keep saturated with weather components in adjoining areas in the city area. Although, during the rainy season, Chittagong city was affected by water logging due to low-lying topography and tidal fluctuation. Some wards were flooded and this water rapidly emission, but for some time it is interrupted by blockage of municipal drain and mismanagement of municipal wastes.

Md. Azadur Rahman Mollick (Director, Department of Environment, Chittagong Metropolitan) claimed that:

> "Urban Water Bodies (UWBs) help city dwellers in different ways, mainly solve of the water problem, saturated microclimate, city area keeps free from water logging, reduce the pressure of groundwater level, Solution of water problem at poor household, water supply for fire service forces and ponds used for fishing purposes."

Mrs. Farjana Parvin (Councilor, 17, 18 and 19 No Wards, Chittagong Metropolitan City) argued that:

> "Urban Water Bodies (UWBs) support of city dwellers in various ways, particularly alternate water sources, keep cool of surrounding environment and place for recreation."

Generally, two factors have accelerated the disappearance of UWBs in Chittagong Metropolitan cities, as internal push factor and external pull factor. Internal push factors included territorial expansions, unplanned urbanization and infrastructural development, the building of new settlements, the violence of real state corporations, and unlawful encroachment and filled-up of water bodies for building commercial and industrial grounds. Besides, the external pull factor was composed of desired jobs and highly lifestyle, getting urban amenities and services, modern life maintenance, and climate displaced person's rehabilitation. For this reason, people enter the city area, and dishonest people build new settlements filled up on UWBs for high rending purposes. For example, many people have already migrated to the urban slums from the coastal zones of Bangladesh due to frequent cyclones, storm surges, river erosion, etc. Besides, the rural poor will in many cases be forced to leave their rural homes and lands for urban areas and slums. This increased urbanization will create even more stress on the different overpopulated cities of Bangladesh, like Dhaka, Chittagong, Khulna other big cities (Barua and Rahman, 2017; Barua and Rahman, 2021).

Institutional responsibility was directly accelerating the natural and man-made causes in this city. During the study, it was heard that land robbers were afraid of the monitoring works and attempts of the Department of Environment (DoE). Because DoE was the lawful authority to manage and protecting of UWBs, they were penalized and filed cases against a few dwellers, who were involved in illegally filled-up water bodies in Chittagong city. Dwellers claimed that maximum water bodies were filled up by the owners of the water bodies. Fragmentation of father or family land, need housing space, the building of housing for rent purposes, an increase of land and getting money off local UWBs owners have been filled-up their capturing land. Although, According to Water Bodies Protection Act, 2000; every water bodies are the land of the government, anybody never filled-up without the permission of the DoE. Besides, 20.36% of respondents argued that 'Real Estate Company' or Developers' second prime institution they have directly involved the filled-up water bodies in Chittagong city. However, NGOs activities, Govt. organizations, political parties, and ownership of land were, directly and indirectly, involved in the filled-up UWBs in Chittagong Metropolitan City (Barua, 2022).

Md. Khorsed *Alam* (Divisional Commissioner, Chittagong division of Bangladesh Government) claimed that:

> "Chittagong Metropolitan city is an industrial hub and commercial capital of Bangladesh. In this city, dwellers are engaged with various types of secondary and tertiary economic activities as a result day by day population gradually increases here. Due to its functional activities, various causes have been accelerated the losses of urban water bodies in Chittagong Metropolitan city, particularly abnormally increase in land prices, industrialization, built multistoried buildings, illegal land capturing, economic pricing, the

influence of politics (the influence of political people, local Mussel man, and powerful elite people) and political violence and so on."

Mrs. Jerina Khan (Professor of Department of Architecture, Chittagong University of Engineering and Technology) argued that:

"Increase in the prices of land, the violence of developers, economic importance and violence of land garbling (recently approximately 40 person land garbling is detected and law and enforcement authority punishment themselves at different orientation) are the main causes for losses of urban water bodies in Chittagong Metropolitan City."

For better management and to ensure urban amenities for city dwellers, different types of autonomous bodies tried to maintain all time. In Chittagong city, different autonomous bodies, particularly Chittagong City Corporation (CCC), Chittagong Development Authority (CDA), Chittagong Water Supply and Sewerage Authority (CWASA), Department of Environment branch of the Metropolitan area. These organizations operate separately with their separate aim and objectives or responsibilities for ensuring safe and happy dwelling life. Management is the first and primary step for mitigation and conservation of UWBs in Chittagong city. Management is an instant step not only for solving existing problems, but also helps future safety or attempts.

Md. Anisur Rahman (Chief Urban Town Planner, Chittagong City Corporation) demanded that:

"Government applying of existing laws particularly "All Playgrounds of Municipal Area, Open Area, Park and natural water reservoir including Municipal Area of City, Divisional Town & District Town, Conservation Act, 2000," strengthening of law and enforcement authority for applying for their role without any bias and political influences, increase awareness program of city dwellers by different advocacy process and the environmental movement at different environmental days, create monitoring cell and punishment should be optimum ways are accelerating the proper management of urban water bodies in Chittagong Metropolitan City."

Mrs. Rebeka Sultana (Member of National Parliament, Chittagong City Area) argued that:

"Cleaning of existing water bodies, elite and political persons should involve the management activities, stimulated of surrounding dwellers for preserving of water bodies by CDA, increase awareness of local people and unused water bodies should be clean and use lime to clear of water is the management of urban water bodies in Chittagong Metropolitan City."

In the book chapter, the authors mentioned that due to losses of water bodies, dwellers are burdened with extra household expenditures. There are four types of consequences that have been faced due to losses of UWBs in Chittagong city. Most consequences are buying water from WASA followed by stopping or hampered local retail shops and tea stalls. Urbanization demands infrastructure development and basic amenities. It results in an increase in impermeable surfaces in the form of pavements and roads that may reduce the natural infiltration of rainwater and increase the surface runoff. Urban development with increased hard paved areas raises the volume of runoff by up to 6 times urban expansion can lead to radical changes like the surface of the urbanized area.

The Management Approach for Urban Water Bodies

The authors highlighted the management of urban water bodies of Chittagong city by focusing on short-term and long-term strategies in Table 16.1. For metropolitan, commercial capital, industrial hub, and recreational city status, Chittagong city dwellers were sufferers of different problems, especially population pressure, unplanned urbanization, infrastructural development, etc. These would accelerate the inadequate supply of urban amenities, particularly the supply of water and electricity during the summer season, mismanagement of solid wastes, waterlogging, and drainage congestion.

Although, the water crisis at the household level was the prime city problem, temperature increase, and waterlogging are equally considered the second prime problem in the whole metropolitan area. However, urban floods, economic losses, and losses of aesthetic views were other problems in the city area. In Chittagong region is situated by tertiary hilly also Chittagong city is covered in this region. Hill, plain land, and low-lying topography were appearing here. Maximum topography was low-lying, especially in southeastern and west-southern regions, and the central part was covered by the hill. Due to reducing the Urban Water Bodies (UWBs), rainwater and surface runoff water didn't hold as a result, this city was affected by waterlogging and urban floods during the rainy season. Furthermore, excessive rainwater accelerated the surface runoff this runoff leads to a huge volume of water and bear debris materials. Due to losses of water bodies, this debris is deposited on drains and low-lying areas. It also contributed to the urban flood and waterlogging during the rainy season in Chittagong city. According to the IPCC report, global temperature has been gradually increasing, and Chittagong city temperature also increases due to evapotranspiration and radiation of heat from unplanned urbanization and infrastructural development, filled-up of urban water bodies have been led to tamper of city temperature.

Table 16.1. Management of Urban Water Bodies (UWBs)

Types of Duration	Management
Short term management	Regular monitoring of the water bodies
	By using properly and not dumping the waste
	To prevent the overexploitation of water bodies
	Stopping the illegal encroachment of ponds and pond wall
	Create strong authority initiatives
	Waste dumps totally prohibited on water bodies
	Proper use & clear of water bodies
	Use pond water after boiling
Long term management	Digging, clearing & authorizing
	Authorization of water bodies
	Building awareness of city dwellers by posting miking, wall writing, and distribution of leaflets
	Build pond wall
	Community movement (local elite, lawyers, representatives of urban autonomous bodies, media men, journalists, researchers, and local youth) for saving water bodies.

Md. Shamsul Alam (Chief City Town Planner, Chittagong Development Authority) demanded that:

> "Due to losses of urban water bodies, city dwellers are facing different types of problems like urban flood, waterlogging (losses or broken off the road network, dwellers suffer water-borne disease, losses of time or schedule and academic works), inadequate water during fire hazards in city areas, facing a daily water crisis and economic problems."

Mr. Rezaul Karim Chowdhury (Mayor of Chittagong City Corporation) argued that:

> "Losses of water bodies, city dwellers face household water crisis, temperate surrounding weather components, a huge volume of water withdrawals by

tube well, groundwater table go underway and pond fish sector fully destroyed."

Conservation Approach of Urban Water Bodies

For a better and safer environment, need proper implementation of existing laws, rules, and regulations, strengthening of law and enforcement authority, particularly CCC, CDA, DoE, etc. Besides, increase the awareness level of local people through banner writing, leaflet distribution, and open seminar those local people attended easily and concerning officials visiting the city area. For a healthy city, environmental conservation is a precondition for the future generation. The current field survey reveals the different types of conservation measures for safe UWBs for future dwellers.

Table 2 illustrates the 13 types of conservation measures to save the UWBs in Chittagong Metropolitan city. Among them is defragmentation of the fatherland (water bodies), properly apply of existing law, rules, and regulations for stopping the illegal encroachment of UWBs, building pond walls, and applying the 'Building Code', lawful authority must be visited before authorization of new building and industry were the prime measures for conserve of UWBs in Chittagong Metropolitan city. Although, proper monitoring of law and enforcement authority and awareness of small owners or local owners of UWBs can be conserved water bodies in Chittagong metropolitan city.

Table 16.2. Conservation of Urban Water Bodies

S.L. No	Conservation Strategies
1	Properly apply existing laws, rules, and regulations for stopping the illegal encroachment of UWBs
2	Taking steps from govt. to clean the water bodies
3	Applied 'Building Code', the lawful authority must be visited before authorization of new building and industry
4	Every industry builds an 'Effluent Treatment Plant' (ETP)
5	Protect water bodies from industrial liquid waste
6	Proper management of municipal waste
7	Stopping pollution
8	Build water reserve
9	Build pond wall
10	Dragging bed of water bodies
11	Use chemicals to purify water
12	Defragmentation of the fatherland (water bodies)

Review of Existing Laws, Rules, and Regulation for Protection of UWBs

Bangladesh is a land of rivers/canals and large sources of drinking water. Apart from natural water bodies, this country has numerous artificially created water resources like ponds, lakes, and canals. To eliminate the shortage of drinking water, numerous large numbers of artificial reservoirs had been created here. But in course of time, the natural and artificial water reservoirs had been subjected to pollution, forceful possession, and filling up.

In recent years, the levels of incidence of forceful possession of water bodies have grown in great numbers. The water bodies in this country are being forcefully possessed filled up and polluted both by Govt. and non-Govt. agencies. Although there is no clear-cut legislation for the protection of water bodies under private ownership, there are laws to protect natural reservoirs. But lack of coordination among the responsible department, weak enforcement of laws, limitation of proper planning and monitoring, due to lack of planning and foresight many reservoirs in this land have been destroyed.

Due to the relentless activities of environmentally concerned people and organizations towards conservation of the country's water bodies, the Government has expected to take a widespread program to protect the water bodies. These programs include the creation and amendment of laws, enhancement of authority of institutions, and monitoring and strengthening of enforcement of legislation. Maintenance of strict view of the Apex Court towards conservation of water bodies, enforcement of the existing laws has gathered momentum to a great extent. The application and enforcement of these legislations can play a notable role in the conservation of water reservoirs in the country.

By the National Drinking Water Policy, a National Drinking Water Resources Council has been instituted consisting of representatives from different Ministries. The Council will bear all the responsibilities related to the materialization of the water arrangement program, and the water resources planning commission will look after the secretarial activities of the council. The Government is associated with more than 40 organizations regarding water arrangement issues. There are certain specific laws for the management of drinking water. However, in the laws and policies of other organizations of drinking water, conservation, and management of water bodies have been emphasized (Alam, 2014).

The Bangladesh Environmental Conservation Act passed in 1995, and the accompanying 1997 Rules, are arguably the most important legislative documents for industrial water pollution. The Act is dedicated to the "conservation, improvement of quality standards, and control through mitigation of pollution of the environment" (Environmental Conservation Act, 1995). The 1997 Environment Conservation Rules made by the 1995 Act provide additional guidance for specific components of the Act. The Act is in theory enforced by the DoE, which has responsibility for 1) Coordinating with other authorities or agencies that have relevance to the objectives of the Act; 2) Adopting safety measures and determining abatement measures to prevent accidents that may cause environmental degradation; 3) Advising persons on environmentally sound use, storage, transportation, import and export of hazardous material or its components; 4) Conducting research and assisting other authorities and agencies in conservation and improvement of the environment; 5) Investigating locations, equipment, manufacture or other processes, ingredients, or materials, to ensure improvement of the environment, and control and mitigation of pollution; 6) Collecting, publishing and disseminating information regarding environmental pollution; 7) Advising the Government on manufacturing processes and materials that may cause pollution and 8) Ensuring potable water quality.

The 2000 Environmental Court Act supports the Environmental Conservation Act (1995) and the Environmental Conservation Rules (1997) by providing for the establishment of environmental courts for the trial of offenses relating to environmental pollution. It includes protocols for the establishment of the court and defines the court's jurisdiction, appropriate penalties, powers of search and entry, and procedures for investigation, trial, and appeal. The Environmental Conservation Act, 1995 and the Environmental Court Act 2000 were amended in 2002, and the Environmental Conservation Rules, 1997 were extended to incorporate ambient air quality and exhaust fan vehicles. "All Playgrounds of Municipal Area, Open Area, Park and natural water reservoir including Municipal Area of City, Divisional Town & District Town, Conservation Act, 2000."

Conclusion

From the earliest to the contemporary period urban water bodies were many times neglected, discharge of solid and liquid wastes into it, partly filled up or eliminated for land development, or simply sometimes regulated within a concrete channel to provide rapid discharge for floodwaters. Environmental concerns or aspects of aesthetics were not considered. Due to the deterioration of these urban water bodies, wealth in the form of land value was destroyed. As already stated, any rehabilitation process of an urban water body requires significant cleanup efforts, a major overhaul of the humid land, and the cleaning of the inflowing water. Particularly, the purification of the currently untreated sewage is a major concern and requires a significant effort. The rehabilitation of smaller watercourses up to now has not been a major topic. Besides, city dwellers have profited multi-dimensions benefits of this water body in this city.

Unplanned development has caused illegal encroachment and filling of vacant lands, unplanned urbanization and housing, sewage problem, local drainage congestion, industrial and household waste disposal problems, and scarcity of urban amenities, particularly supply water and so on. But most of the illegal activities originated from a neglectful attitude of law and enforcement authorities which directly accelerated every illegal activity in every area. So, planned urban development is not only a precondition of territorial expansion of urban areas but also deliberated planned residential, commercial, industrial, and recreational areas that are safe and conserve the urban environment, particularly urban water bodies.

Recommendations

In this book chapter, the authors recommended necessary conservation and management of urban water bodies as alternate water sources for city dwellers:

1. To ensure coordination among different urban self-governing bodies (i.e., Police commissioner, DoE, DoF, CDA, CCC, CWASA, and Land office) in the metropolitan area. Such coordination should be under the leadership of the ‘City Mayor’. Every concerned public agency, private sector, NGOs, and others should be involved in coordination initiatives.
2. Coordination committee created a vigilance team, and every ward commissioner is considered a member of this committee. This committee with law and enforcement authority must be removed from every encroachment on water bodies and periodic monitoring of water bodies. Besides, political parties should involve with this

committee and be aware of involving themselves in conflicts of interest such as violating laws and land grabbing.

3. Master Plan (MP), Detail Area Plan (DAP), and Drainage Network Plan (DNT) must be formulated by taking into consideration the context, problems, and prospects of the existing condition of a metropolitan city. These plans should contain clear guidelines about structural development, territorial expansion, urban open and vacant places, urban water bodies, land use and land zoning, installment of municipal drainage and sewerage lines, and ensuring urban amenities and services for the city dwellers.
4. For strengthening of Law and Enforcement conditions in the metropolitan area, Chittagong City Corporation (CCC), DoE, and Police Commissioner should be ensured law and enforcement at every level. Because Bangladesh has various effective Laws and Acts, particularly *'Local Government (City Corporation), Act 2009'*, *'Environment Court Rules-2000'*, *'The Rivers, Flood Plains and Water Bodies Conservation, Development and Reclamation, and Filling Restriction Act 2005'*. *'Jaladhar Songrokkhon Aain, 2000'*. *'All Playgrounds of Municipal Area, Open Area, Park and natural water reservoir including Municipal Area of City, Divisional Town & District Town, Conservation Act, 2000'* for conservation and management or protection of urban water bodies in urban areas of Bangladesh. This Law and Acts are empowered and do forbid the filling of urban water bodies in every type of urban area of Bangladesh.
5. *'The Building Construction Act, 1952 & 1996'* must be ensured at every level. In essential cases, professional planners, engineers, architects, and public representatives concerning agencies must be consulted before providing permission for infrastructural development. It not only stopped illegal encroachment of urban water bodies for high-rise apartment or residential and commercial purposes but also unplanned industrial development of those polluted surface water bodies and the surrounding environment.
6. Law and enforcement authorities must control the illegal violence of the 'Real Estate Agency' by applying *'The Real Estate Development and Control Act, 2010'* because from 1995 to 2010 Real Estate Agency is the main reason for filled-up water bodies in Chittagong Metropolitan City.

7. For conservation of water bodies, every water body should be declared a 'Protected Area', and ward commissioners should arrange swimming competitions on surrounding dwellers and DoE, CCC, or CDA escape of native fish species every year of every water body. Ward commissioner chose a security person from the local community, this person updated information and monitoring every water body in his area, and wards commissioner arrange the allowance of a security person by selling fish from water bodies.

Finally, city dwellers should be familiar with 'rainwater harvesting' during the rainy season because the average rainfall is 2320 millimeters for nearly four decades in Bangladesh. Chittagong city is situated in the north-south region and rainfall-dominated area of Bangladesh. Heavy rainfall occurred from May to October for a total of six months in these regions. During those months (May to October) city dwellers can hold rainwater for household purposes.

References

Astel, A, Tsakovski, S, Barbieri, P and Simeonov, V. 2017. Comparison of self-organizing maps classification approach with cluster and principal components analysis for large environmental data sets'. *Water Research,* 41 (3), 4566–4578.

Alam, M. E., Chowdhury, M. A. T., and Akbar, M. T. 2013. Aspect of socio-environmental problems resulting from hill cutting in the city of Chittagong Bangladesh: Local people's perspectives. *Pakistan Journal of Social Sciences,* 13(8), 1048-1052.

Barua, P. 2022. Urban water quality index of the commercial capital city of Bangladesh. In: Chatterjee et al. (Ed.) *Advances in Urbanism, Smart Cities, and Sustainability.* CRC Press and Tailors ad Francis Group. pp. 181-220.

Barua, P. and Rahman, S. H. 2020. Relationship between climate change vulnerability and adaptation practices in South-Eastern Coast of Bangladesh. *Asian Profile,* 49(2), 45-60.

Barua, P., Rahman, S. H., and Molla, M. H. 2017. Sustainable adaptation for resolving climate displacement issues of south eastern islands in Bangladesh," *International Journal of Climate Change Strategies and Management,* 15 (2), 60-80.

Barua, P., Rahman, S. H., Rahman, I. K., and Barua, S. 2021. Climate change and its impact on fishermen community of South-Eastern Coast of Bangladesh. *Water Conservation and Management,* 4(2), 55-75.

Hasan, A. B., Kabir, S., Reza, A. H. M., Zaman, M. N., Ahsan, M. A., Akbor, M. A., & Rashid, M. M. 2013. Trace Metals pollution in seawater and groundwater in the ship breaking areas of Sitakund upazila, Chittagong, Bangladesh. *Marine Pollution Bulletin,* 71(1–2), 317–324.

Hassan, M. M. and Nazem, M. N. I. (2015). Examination of land use/land cover changes, urban growth dynamics, and environmental sustainability in Chittagong city, Bangladesh. *Environment Development and Sustainability,* 45(5), 50-65.

Kaufmann, R. K., Seto, K. C., Schneider, A., Liu, Z., Zhou, L., and Wang, W. 2017. Climate response to rapid urban growth: Evidence of a human-induced precipitation deficit. *Journal of Climate,* 20(10), 2299–2306.

Rahman, I. M. M., Islam, M. M., Hossain, M. M., Hossin, M. S., Begum., Z. A., Chowdhury, D. A., Chakraborty, M. K., Rahman, M. A., Nazimuddin, M. and Hasegawa, H. 2011. Stagnant surface water bodies (SSWBs) as an alternative water resource for the Chittagong metropolitan area of Bangladesh: physicochemical characterization in terms of water quality indices. *Environmental Monitoring and Assessment,* 173 (1), 669–684.

Rahman, M. S. 2012. Quantifying urban form-compactness versus sprawl: An Analysis of Chittagong and Sylhet metropolitan cities. *Jahangirnagar University Planning Review*, 10(3): 39-50.

Reza, T., Bakar, M. A., Islam, S and Bhuiyan, H. R. 2017. Pollution level and monthly fluctuation of pollution - A study of wastewater of Chittagong city, Bangladesh. *Electronic Journal of Environmental, Agricultural and Food Chemistry,* 9 (1) : 234-238.

About the Editors

Saeid Eslamian

Prof. Saeid Eslamian has graduated at PhD level from Department of Civil Engineering, University of New South Wales, Australia in 1995. He is now a Full Professor of Hydrology and Water Resources Sustainability at Isfahan University of Technology, Director of Excellence Center in Risk Analysis and Natural Hazards and Chief of Specialized Group on Water Resources Policy and Management, Research Center of Water and Wastewater.

Formerly, he was a Visiting Professor at Princeton University, United States, University of ETH Zurich, Switzerland, and McGill University, Montreal, Quebec, Canada. Recently, he has been awarded as one of the World's Top 2% Scientists from Stanford University and awarded High-level Jury Member and Invited Dialogue Person for the related organizations of United Nations.

Prof. Eslamian is the Founder and Chief Editor of *International Journal of Hydrological Science and Technology* (Inderscience) and *Water Productivity Journal*. Prof. Eslamian is also the editorial board member and reviewer of more than 150 Web of Science (ISI) Journals. He has contributed to more than 1K publications in journals, books, and technical reports including Elsevier, Springer, Wiley, Taylor and Francis and IWA.

Mir Bintul Huda

Mir Bintul Huda has a Doctorate in Civil Engineering. She has done her M. Tech in Water Resources Engineering, Bachelors in Agricultural Engineering with majors in Soil and Water Conservation Engineering and AMIE (Civil Engineering). She is a corporate member of IEI, India. Presently she works as a Consultant at Water Resources Management Centre, National Institute of Technology, Jammu & Kashmir. In addition to the research work, she conducts lectures for doctoral candidates. She has specialization in Soil and Water Resources Engineering, Hydraulic Structures, River Engineering and Climate Change Studies. She has numerous publications in reputed peer reviewed international journals and books and is a part of several academic houses and organizations.

Faezeh Eslamian

Dr. Faezeh Eslamian holds a Ph.D. Degree in Bioresource Engineering from McGill University, Canada as well as a Bachelors and Masters degree in Civil and Environmental engineering. She has more than 10 years of experience in managing various multidisciplinary projects including soil and groundwater characterization and remediation, hydrogeology, water, and waste water treatment and numerical modeling.

Dr. Faezeh has more than 40 publications including journal papers, book chapter as well as handbooks.

Index

S

T

U

V

W

Z